T0142855

Studies in Computational Intelligence

Volume 678

Series editor

Janusz Kacprzyk, Polish Academy of Sciences, Warsaw, Poland
e-mail: kacprzyk@ibspan.waw.pl

About this Series

The series "Studies in Computational Intelligence" (SCI) publishes new developments and advances in the various areas of computational intelligence—quickly and with a high quality. The intent is to cover the theory, applications, and design methods of computational intelligence, as embedded in the fields of engineering, computer science, physics and life sciences, as well as the methodologies behind them. The series contains monographs, lecture notes and edited volumes in computational intelligence spanning the areas of neural networks, connectionist systems, genetic algorithms, evolutionary computation, artificial intelligence, cellular automata, self-organizing systems, soft computing, fuzzy systems, and hybrid intelligent systems. Of particular value to both the contributors and the readership are the short publication timeframe and the worldwide distribution, which enable both wide and rapid dissemination of research output.

More information about this series at http://www.springer.com/series/7092

Costin Badica · Amal El Fallah Seghrouchni
Aurélie Beynier · David Camacho
Cédric Herpson · Koen Hindriks
Paulo Novais
Editors

Intelligent Distributed Computing X

Proceeding of the 10th International
Symposium on Intelligent Distributed
Computing – IDC 2016, Paris, France,
October 10–12 2016

 Springer

Editors
Costin Badica
Faculty of Automatics, Computer Science
   and Electronics
University of Craiova
Craiova
Romania

Amal El Fallah Seghrouchni
Sorbonne Universités
UPMC Univ Paris 06, CNRS,
   LIP6 UMR 7606
Paris
France

Aurélie Beynier
Sorbonne Universités
UPMC Univ Paris 06, CNRS,
   LIP6 UMR 7606
Paris
France

David Camacho
Universidad Autonoma de Madrid
C. Francisco Tomas y Valiente, 11
Madrid
Spain

Cédric Herpson
Sorbonne Universités
UPMC Univ Paris 06, CNRS,
   LIP6 UMR 7606
Paris
France

Koen Hindriks
Faculty of EEMCS
Delft University of Technology
Delft, Zuid-Holland
The Netherlands

Paulo Novais
Universidade do Minho
Campus of Gualtar
Braga
Portugal

ISSN 1860-949X   ISSN 1860-9503   (electronic)
Studies in Computational Intelligence
ISBN 978-3-319-84024-6   ISBN 978-3-319-48829-5   (eBook)
DOI 10.1007/978-3-319-48829-5

Printed on acid-free paper

This Springer imprint is published by Springer Nature
The registered company is Springer International Publishing AG
The registered company address is: Gewerbestrasse 11, 6330 Cham, Switzerland

Preface

Intelligent Distributed Computing emerged as the result of the fusion and cross-fertilization of ideas and research results in Intelligent Computing and Distributed Computing. Its main outcome was the development of the new generation of intelligent distributed systems, by combining methods and technology from classical artificial intelligence, computational intelligence, multi-agent-systems, and distributed systems.

The 10th Intelligent Distributed Computing IDC2016 continues the tradition of the IDC Symposium Series that started 10 years ago as an initiative of two research groups from:

1. Systems Research Institute, Polish Academy of Sciences, Warsaw, Poland
2. Software Engineering Department, University of Craiova, Craiova, Romania

The IDC Symposium welcomes submissions of original papers on all aspects of intelligent distributed computing ranging from concepts and theoretical developments to advanced technologies and innovative applications. The symposium aims to bring together researchers and practitioners involved in all aspects of Intelligent Distributed Computing. IDC is interested in works that are relevant for both Distributed Computing and Intelligent Computing, with scientific merit in these areas.

This volume contains the proceedings of the 10th International Symposium on Intelligent Distributed Computing, IDC2016. The symposium was hosted by the Laboratoire d'Informatique de Paris 6 from the University Pierre and Marie Curie, in Paris, France, between the 10th and the 12th of October, 2016.

The IDC2016 event comprised the main conference organized in eight sessions: Dynamic Systems, Internet of Things, Security, Space-Based Coordination, Behavioral Analysis, Optimization, Data Management and IC-Smart.

The proceedings book contains contributions, with 23 regular papers selected from a total of 38 received submissions from 18 countries (counting the country of each co-author for each paper submitted). Each submission was carefully reviewed by at least three members of the Program Committee. Acceptance and publication were judged based on the relevance to the symposium topics, clarity of presentation, originality and accuracy of results, and proposed solutions. The acceptance rates were 60%, counting only regular papers. The contributions published in this book address many topics related to theory and applications of intelligent distributed computing including: cloud computing, P2P networks, agent-based distributed simulation, ambient agents, smart and context-driven environments, Internet of Things, network security, mobile computing, Unmanned Vehicles, augmented physical reality, swarm computing, team and social computing, constraints and optimization, and information fusion.

We would like to thank Janusz Kacprzyk, editor of Studies in Computational Intelligence series and member of the Steering Committee, for his continuous support and encouragement for the development of the IDC Symposium Series. Also, we would like to thank the IDC2016 Program Committee members for their work in promoting the event and refereeing submissions and also to all colleagues who submitted their work to this event.

We deeply appreciate the efforts of our invited speakers Pr. Serge Haddad from ENS Cachan, France, and Pr. Carlos COTTA from Universidad de Mlaga, Spain, and thank them for their interesting lectures.

A special thanks also go to organizers of the special session IC-Smart : Amal El Fallah Seghrouchni and Kenji Tei. Finally, we appreciate the efforts of local organizers on behalf of the Laboratoire d'Informatique de Paris 6 (LIP6) from the University Pierre and Marie Curie, Sorbonnes Universities, in Paris, France, for hosting and organizing this event.

Craiova	Costin Badica
Paris	Amal El Fallah Seghrouchni
Paris	Aurelie Beynier
Madrid	David Camacho
Paris	Cedric Herpson
Zuid-Holland	Koen Hindriks
Braga	Paulo Novais

July 2016

Organization

Organizer

MultiAgent System team (SMA)
Laboratoire d'Informatique de Paris 6 (LIP6)
Sorbonne Universités, UPMC Univ Paris 06, France

General Chairs

Amal El Fallah Seghrouchni	LIP6 - UPMC Sorbonne Universités, France
Costin Badica	University of Craiova, Romania

Program Committee Chairs

Aurélie Beynier	UPMC Sorbonne Universités, LIP6, France
David Camacho	Universidad Autonoma de Madrid, Spain
Koen Hindriks	Delft Robotics Institute, The Netherlands
Paulo Novais	University of Minho, Portugal

Invited Speakers

Carlos Cotta	Universidad de Malaga, Spain
Serge Haddad	ENS Cachan, France

Program Comittee

Ajith Abraham	Machine Intelligence Research Labs (MIR Labs)
Salvador Abreu	JFLI-CNRS / LISP / CRI, University of Evora
Amparo Alonso-Betanzos	University of A Corua
Ricardo Anacleto	ISEP
Cesar Analide	University of Minho
Razvan Andonie	Central Washington University
Javier Bajo	.Universidad Politécnica de Madrid
Nick Bassiliades	Aristotle University of Thessaloniki
David Bednrek	Charles University Prague
Doina Bein	California State University, Fullerton
Gema Bello Orgaz	Universidad Autonoma de Madrid
Nik Bessis	Edge Hill University
Lars Braubach	University of Hamburg
Dumitru Dan Burdescu	University of Craiova
Giacomo Cabri	Universit di Modena e Reggio Emilia
Davide Carneiro	Universidade do Minho
Andre Carvalho	USP

Jose Carlos Castillo Montoya	Universidad Carlos III de Madrid
Jen-Yao Chung	IBM
Dorian Cojocaru	University of Craiova
Rem Collier	UCD
Phan Cong-Vinh	Nguyen Tat Thanh University
Lus Correia	University of Lisbon
Ängelo Costa	Universidade do Minho
Paul Davidsson	Malm University
Javier Del Ser	Tecnalia Resaerch & Innovation
Giuseppe Di Fatta	University of Reading
Amal El Fallah Seghrouchni	UPMC Sorbonne Universités, LIP6, France
Vadim Ermolayev	Zaporozhye National Univ.
Antonio Fernndez-Caballero	Universidad de Castilla-La Mancha
Adina Magda Florea	University Politehnica of Bucharest, AI-MAS Laboratory
Giancarlo Fortino	University of Calabria
Maria Ganzha	University of Gdask
Antonio Gonzalez-Pardo	Universidad Autonoma de Madrid
Bertha Guijarro-Berdias	University of A Corua
Marjan Gusev	UKIM University St Cyril and Methodius
Adnan Hashmi	University of Lahore
Cédric Herpson	UPMC Sorbonne Universités, LIP6, France
Dosam Hwang	Yeungnam University
Barna Laszlo Iantovics	Petru Maior University of Tg. Mures
Fuyuki Ishikawa	National Institute of Informatics
Mirjana Ivanovic	University of Novi Sad, Faculty of Sciences
Vicente Julian	GTI-IA DSIC UPV
Jason Jung	Chung-Ang University
Igor Kotenko	St. Petersburg Institute for Informatics and Automation
Dariusz Krol	Wrocaw University of Technology
Florin Leon	Technical University "Gheorghe Asachi" of Iasi
Alessandro Longheu	DIEEI - University of Catania
José Machado	University of Minho, Computer Science and Technology Centre
Ana Madureira	Departamento de Engenharia Informtica
Giuseppe Mangioni	DIEEI - University of Catania
Goreti Marreiros	ISEP/IPP-GECAD
Ester Martinez-Martin	Universitat Jaume I
Viviana Mascardi	Department of Informatics, University of Genova
Ficco Massimo	Second University of Naples (SUN)
Héctor Menéndez	University Autonoma of Madrid

John-Jules Meyer	Utrecht University
Paulo Moura Oliveira	UTAD University
Grzegorz J Nalepa	AGH University of Science and Technology
Jose Neves	Universidade do Minho
David Obdrzalek	Charles University in Prague
Andrea Omicini	Alma Mater StudiorumUniversit di Bologna
Fernando Otero	University of Kent
Juan Pavn	Universidad Complutense de Madrid
Pawel Pawlewski	Poznan University of Technology
Stefan-Gheorghe Pentiuc	University Stefan cel Mare Suceava
Antonio Pereira	Escola Superior de Tecnologia e Gesto do IPLeiria
Dana Petcu	West University of Timisoara
Florin Pop	University Politehnica of Bucharest
Antonio Portilla-Figueras	Universidad de Alcala
Maria Potop-Butucaru	UPMC Sorbonne Universités, LIP6, Paris
Radu-Emil Precup	Politehnica University of Timisoara
Maria D. R-Moreno	Universidad de Alcala
Shahram Rahimi	Southern Illinois University
Alessandro Ricci	University of Bologna
Joel J.P.C. Rodrigues	Instituto de Telecomunicaes, University of Beira Interior
Domenico Rosaci	DIMET Department, University Mediterranea of Reggio Calabria
Sancho Salcedo-Sanz	Universidad de Alcal
Corrado Santoro	University of Catania - Dipartimento di Matematica e Informatica
Ichiro Satoh	National Institute of Informatics
Weiming Shen	National Research Council
Fbio Silva	Universidade do Minho
Safeeullah Soomro	Indus University
Giandomenico Spezzano	CNR-ICAR and University of Calabria
Stanimir Stoyanov	University of Plovdiv "Paisii Hilendarski"
Anna Toporkova	National Research University Higher School of Economics
Rainer Unland	University of Duisburg-Essen, ICB
Salvatore Venticinque	Seconda Universit di Napoli
Lucian Vintan	"Lucian Blaga" University of Sibiu
Martijn Warnier	Delft University of Technology
Michal Wozniak	Wroclaw University of Technology
Jakub Yaghob	Charles University in Prague
Filip Zavoral	Charles University in Prague

Organizing Committee

Aurélie Beynier   UPMC Sorbonne Universités, LIP6, France
Amal El Fallah Seghrouchni   UPMC Sorbonne Universités, LIP6, France
Cédric Herpson   UPMC Sorbonne Universités, LIP6, France

Sponsoring Institutions

Sorbonne Universités, UPMC Univ Paris 06
Laboratoire d'Informatique de Paris 6 (LIP6)
Faculté d'Ingénierie, UFR 919, UPMC

Table of Contents

IV   Space-Based Coordination

V   Behavioral Analysis

VI   Optimization

VII   Data Management

Part I

Dynamic Systems

Adaptive Scaling Up/Down for Elastic Clouds

Ichiro Satoh

National Institute of Informatics
2-1-2 Hitotsubashi, Chiyoda-ku, Tokyo, Japan
ichiro@nii.ac.jp

Abstract. An approach for adapting distributed applications in response to changes in user requirements and resource availability is presented. The notion of *elasticity* enables capabilities and resources to be dynamically provisioned and released. However, existing applications do not inherently support *elastic* capabilities and resources. To solve this problem, we propose two novel functions: *dynamic deployment of components* and *dividing and merging components*. The former enables components to relocate themselves at new servers when provisioning the servers and at remaining servers when deprovisioning servers, while the latter enables the states of components to be divided, passed to other components, and merged with other components in accordance with user-defined functions. We constructed a middleware system for Java-based general-purpose software components with the two functions because they are useful to adapt applications to elasticity in cloud computing. The proposed system is useful because it enables applications be operated with elastic capabilities and resources in cloud computing.

1 Introduction

Elasticity in cloud computing was originally defined in physics as a material property with the capability of returning to its original state after a deformation. The concept of elasticity has been applied to computing and is commonly considered to be one of the central attributes of cloud computing. For example, the NIST definition of cloud computing [10] states that capabilities can be elastically provisioned and released, in some cases automatically, to scale rapidly outward and inward in accordance with demand. To the consumer, the capabilities available for provisioning often appear to be unlimited and can be appropriated in any quantity at any time.

However, the conventional design and development of application software are not able to adapt themselves to elastically provisioning and deprovisioning resources in cloud computing. Furthermore, it is difficult to deprive parts of the computational resources that such applications have already used. There have been a few attempts to solve this problem. For example, Mesos [4] is a platform for sharing commodity clusters between distributed data processing frameworks such as Hadoop and Spark. These frameworks themselves are elastic in the sense that they have the ability to scale their resources up or down, i.e., they can start using resources as soon as applications want to acquire the resources or release the resources as soon as the applications do not need them.

3

C. Badica et al. (eds.), *Intelligent Distributed Computing X*,
Studies in Computational Intelligence 678, DOI 10.1007/978-3-319-48829-5_1

We assume that applications are running on dynamic distributed systems, including cloud computing environments, in the sense that computational resources available from the applications may be dynamically changed due to elasticity. We propose a framework for enabling distributed applications to be adapted to changes in their available resources on elastic distributed systems as much as possible. The key ideas behind the framework are the duplication and migration of running software components and the integration of multiple same components into single components. To adapt distributed applications, which consist of software components, to elasticity in cloud computing, the framework divides applications into some of the components and deploys the components at servers, which are provisioned, and merges the components running at servers, which are deprovisioned, into other components running at other available servers. We are constructing a middleware system for adapting general-purpose software components to changes in elastic cloud computing.

2 Related Work

Cloud computing environments allow for novel ways of efficient execution and management of complex distributed systems, such as elastic resource provisioning and global distribution of application components. Resource allocation management has been studied for several decades in various contexts in distributed systems, including cloud computing. We focus here on only the most relevant work in the context of large-scale server clusters and cloud computing in distributed systems. Several recent studies have analyzed cluster traces from Yahoo!, Google, and Facebook and illustrate the challenges of scale and heterogeneity inherent in these modern data centers and workloads. Mesos [4] splits the resource management and placement functions between a central resource manager and multiple data processing frameworks such as Hadoop and Spark by using an offer-based mechanism. Resource allocation is performed in a central kernel and master-slave architecture with a two-level scheduling system. With Mesos, reclaim of resources is handled for unallocated capacity that is given to a framework. The Google Borg system [11] is an example of a monolithic scheduler that supports both batch jobs and long-running services. It provides a single RPC interface to support both types of workload. Each Borg cluster consists of multiple cells, and it scales by distributing the master functions among multiple processes and using multi-threading. YARN [14] is a Hadoop-centric cluster manager. Each application has a manager that negotiates for the resources it needs with a central resource manager. These systems assume the execution of particular applications, e.g., Hadoop and Spark, or can assign resources to their applications before the applications start. In contrast, our framework enables running applications to adapt themselves to changes in their available resources.

Several academic and commercial projects have explored attempts to create autoscaling applications. Most of them have used static mechanisms in the sense that they are based on models to be defined and tuned at design time. For example, Tamura et al. [13] proposed an approach to identify system viability zones that are defined as states in which the system operation is not compromised and to verify whether the current available resources can satisfy the validation at the development of the applications. The variety of available resources with different characteristics and costs, variability and unpredictability of workload conditions, and different effects of various configurations of resource allocations make the problem extremely hard if not impossible to solve algorithmically at design time.

Reconfiguration of software systems at runtime to achieve specific goals has been studied by several researchers. For example, Jaeger et al. [6] introduced the notion of self-organization to an object request broker and a publish / subscribe system. Lymberopoulos et al. [9] proposed a specification for adaptations based on their policy specification, *Ponder* [1], but it was aimed at specifying management and security policies rather than application-specific processing and did not support the mobility of components. Lupu and Sloman [8] described typical conflicts between multiple adaptations based on the Ponder language. Garlan et al. [3] presented a framework called Rainbow that provided a language for specifying self-adaptation. The framework supported adaptive connections between operators of components that might be running on different computers. They intended to adapt coordinations between existing software components to changes in distributed systems, instead of increasing or decreasing the number of components.

Most existing attempts have been aimed at provisioning of resources, e.g., the work of Sharman at al. [12]. Therefore, there have been a few attempts to adapt applications to deprovisioned resources. Nevertheless, they explicitly or implicitly assume that their target applications are initially constructed on the basis of master-slave and redundant architectures. Several academic and commercial systems tried introducing *live-migration* of virtual machines (VMs) into their systems, but they could not merge between applications, because they were running on different VMs.[1] Jung et al.[7] have focused on controllers that take into account the costs of system adaptation actions considering both the applications (e.g., the horizontal scaling) and the infrastructure (e.g., the live migration of virtual machines and virtual machine CPU allocation) concerns. Thus, they differ from most cloud providers, which maintain a separation of concerns, hiding infrastructural control decisions from cloud clients.

3 Approach

As mentioned in the first section, *elasticity*, which is one of the most important features of cloud computing, is the degree to which a system is able to adapt to workload changes by provisioning and de-provisioning resources in an autonomic manner. Applications need to adapt themselves to changes in their available resources due to elasticity.

3.1 Requirements

We will propose a framework to adapt applications to the provisioning and deprovisioning of servers, which may be running on physical or virtual machines, and software containers, such as Docker, by providing an additional layer of abstraction and automation of virtualization. Our framework assumes that each application consists of one or

---

[1] Unlike private cloud environments, most commercial ones do not support live migrations because they tend to need wide-band and low-latency networks between servers.

more software components that may be running on different computers. It has five requirements.

- *Supports elasticity:* Elasticity allows applications to use more resources when needed and fall back afterwards. Therefore, applications need to be adapted to dynamically increasing and decreasing their available resources.
- *Separation of concerns:* All software components should be defined independently of our adaptation mechanism as much as possible. This will enable developers to concentrate on their own application-specific processing.
- *Self-adaptation:* Distributed systems essentially lack a global view due to communication latency between computers. Software components, which may be running on different computers, need to coordinate themselves to support their applications with partial knowledge about other computers.
- *Non-centralized management:* There is no central entity to control and coordinate computers. Our adaptation should be managed without any centralized management so that we can avoid any single points of failures and performance bottlenecks to ensure reliability and scalability.
- *General purpose:* There are various applications running on a variety of distributed systems. Therefore, the framework should be implemented as a practical middleware system to support general-purpose applications.

We assume that, before the existence of deprovisioning servers, the target cloud computing environment can notify servers about the deprovisioning after a certain time. Cloud computing environments can be classified into three types: Infrastructure as a Service (IaaS), Platform as a Service (PaaS), and Software as a Service (SaaS). The framework is intended to be used in the second and third, but as much as possible it does not distinguish between the two.

3.2 Adaptation for elasticity in cloud computing

To adapt applications to changes in their available resources due to elasticity, the framework adapts the applications to provisioning and de-provisioning resources (Fig. 1).

- *Adaptation to provisioning resources* When provisioning servers, if a particular component is busy and the servers can satisfy the requirement of that component, the framework divides the component into two components and deploys one of them at the servers, where the divided components have the same programs but their internal data can be replicated or divided in accordance with application-specific data divisions.
- *Adaptation to deprovisioning resources* When deprovisioning servers, components running on the servers are relocated to other servers that can satisfy the requirements of the components. If other components whose programs are the same as the former components co-exist on the latter servers, the framework instructs the deployed components to be merged to the original components.

The first and second adaptations need to deploy components at different computers. Our framework introduces mobile agent technology. When migrating and duplicating components, their internal states stored in their heap areas are transmitted to their destinations and are replicated at their clones.

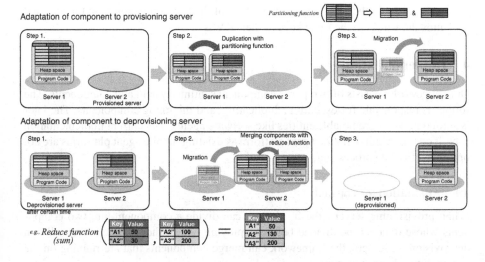

Fig. 1. Adaptation to (de)provisioning servers

3.3 Data stores for dividing and merging components

The framework provides another data store for dividing and merging components. To do this, it introduces two notions: *key-value store* (KVS) and *reduce* functions of the *MapReduce* processing. The KVS offers a range of simple functions for manipulation of unstructured data objects, called *values*, each of which is identified by a unique *key*. Such a KVS is implemented as an array of *key* and *value* pairs. Our framework provides KVSs for components so that each component can maintain its internal state in its KVS. Our KVSs are used to pass the internal data of components to other components and to merge the internal data of components into their unified data. The framework also provides a mechanism to divide and merge components with their internal states stored at KVSs by using *MapReduce* processing. *MapReduce* is a most typical modern computing models for processing large data sets in distributed systems. It was originally studied by Google [2] and inspired by the *map* and *reduce* functions commonly used in parallel list processing (LISP) and functional programming paradigms.

- *Component division* Each duplicated component can inherit partial or all data stored in its original component in accordance with user-defined *partitioning* functions, where each function map of each item of data in its original component's KVS is stored in either the original component's KVS or the duplicated component's KVS without any redundancy.
- *Component fusion* When unifying two components that generated from the same programs into a single component, the data stored in the KVSs of the two components are merged by using user-defined *reduce* functions. These functions are similar to the *reduce* functions of MapReduce processing. Each of our *reduce* functions processes two values of the same keys and then maps the results to the entries of the keys. Figure 1 shows two examples of *reduce* functions. The first concate-

nates values in the same keys of the KVSs of the two components, and the second sums the values in the same keys of their KVSs.

4 Implementation

Our framework consists of two parts: component runtime system and components. The former is responsible for executing, duplicating, and migrating components. The later is autonomous programmable entities like software agents. The current implementation is built on our original mobile agent platform as existing mobile agent platforms are not optimized for data processing.

4.1 Adaptation for elasticity

When provisioning servers, the framework can divide a component into two components whose data can be divided before deploying one of them at the servers. When deprovisioning servers, the framework can merge components that are running on the servers into other components.

Dividing component When dividing a component into two, the framework has two approaches for sharing between the states of the original and clone components.

- *Sharing data in heap space* Each runtime system makes one or more copies of components. The runtime system can store the states of each agent in heap space in addition to the codes of the agent in a bit-stream formed in Java's JAR file format, which can support digital signatures for authentication. The current system basically uses the Java object serialization package for marshalling agents. The package does not support the capturing of stack frames of threads. Instead, when an agent is duplicated, the runtime system issues events to it to invoke their specified methods, which should be executed before it is duplicated, and it then suspends their active threads.
- *Sharing data in KVS* When dividing a component into two components, the KVS inside the former is divided into two KVSs in accordance with user-defined partitioning functions in addition to built-in functions, and the divided KVSs are maintained inside the latter. Partitioning functions are responsible for dividing the intermediate key space and assigning intermediate key-value pairs to the original and duplicated components. In other words, the partition functions specify the components to which an intermediate key-value pair must be copied. KVSs are constructed as in-memory storage to exchange data between components. It provides tree-structured KVSs inside components. In the current implementation, each KVS in each data processing agent is implemented as a hash table whose keys, given as pairs of arbitrary string values, and values are byte array data, and it is carried with its agent between nodes,

where a default partitioning function is provided that uses hashing. This tends to result in fairly well-balanced partitions. The simplest partitioning functions involve computing the hash value of the key and then taking the mod of that value using the number of the original and duplicated components.

Merging components The framework provides a mechanism to merge the data stored in the KVSs of different components instead of the data stored inside their heap spaces. Like the *reduce* of MapReduce processing, the framework enables us to define a *reduce* function that merges all intermediate values associated with the same intermediate key. When merging two components, the framework can discard the states of their heap spaces or keep the state of the heap space of one of them. Instead, the data stored in the KVSs of different components can be shared. A *reduce* function is applied to all values associated with the same intermediate key to generate output key-value pairs. The framework can merge more than two components at the same computers because components can migrate to the computers that execute co-components that the former wants to merge to.

5 Evaluation

We outline our current implementation. A prototype implementation of this framework was constructed with Java Developer Kit (JDK) version 1.7 or later. The implementation enabled graphical user interfaces to operate the mobile agents. Although the current implementation was not constructed for performance, we evaluated the performance of our framework with CoreOS, which is a lightweight operating system based on Linux with JDK version 1.8 with Docker, which is software-based environment that automates the deployment of applications inside software containers by providing an additional layer of abstraction and automation of operating-system-level virtualization on Linux, on Amazon EC2. For each dimension of the adaptation process with respect to a specific resource type, elasticity captures the following core aspects of the adaptation:

- *Adaptation speed at provisioning servers* The speed of scaling up is defined as the time it takes to switch from provisioning of servers by the underlying system, e.g., cloud computing environment.
- *Adaptation speed at deprovisioning servers* The speed of scaling down is defined as the time it takes to switch from deprovisioning of servers by the underlying system, e.g., cloud computing environment.

The speed of scaling up/down does not correspond directly to the technical resource provisioning/deprovisioning time. Table 1 shows the basic performance. The component was simple and consisted of basic callback methods. The cost included that of invoking two callback methods. The cost of component migration included that of opening TCP transmission, marshaling the agents, migrating the agents from their source computers to their destination computers, unmarshalling the components, and verifying security.

Table 1. Basic operation performance

	Latency (ms)
Duplicating component	10
Merging component	8
Migrating component between two servers	32

Figure 2 shows the speed of the number of divided and merged components at provisioning and deprovisioning servers. The experiment provided only one server to run our target component, which was a simple HTTP server (its size was about 100 KB). It added one server every ten seconds until there were eight servers and then removed one server every ten seconds after 80 seconds had passed. The number of components was measured as the average of the numbers in ten experiments. Although elasticity is always considered with respect to one or more resource types, the experiment presented in this paper focuses on computing environments for executing components, e.g., servers. There are two metrics in an adaptation to elastic resources, *scalability* and *efficiency*, where scalability is the ability of the system to sustain increasing workloads by making use of additional resources, and efficiency expresses the amount of resources consumed for processing a given amount of work.

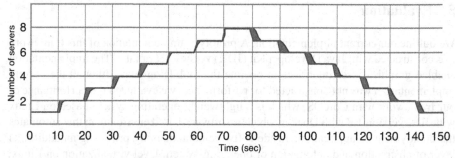

Fig. 2. Number of components at (de)provisioning servers

- $\overline{A}$ is the average time to switch from an underprovisioned state to an optimal or overprovisioned state and corresponds to the average speed of scaling up or scaling down.
- $\overline{U}$ is the average amount of underprovisioned resources during an underprovisioned period. $\sum \overline{U}$ is the accumulated amount of underprovisioned resources and corresponds to the blue areas in Fig. 2.
- $\overline{D}$ is the average amount of overprovisioned resources during an overprovisioned period. $\sum \overline{D}$ is the accumulated amount of underprovisioned resources and corresponds to the red areas in Fig. 2.

The precision of scaling up or down is defined as the absolute deviation of the current amount of allocated resources from the actual resource provisioning or deprovisioning. We define the average precision of scaling up P_u and that of scaling down P_d. The efficiency of scaling up or down is defined as the absolute deviation of the accumulated amount of underprovisioned or overprovisioned resources from the accumulated amount of provisioned or deprovisioned resources, specified as E_U or E_D.

$$P_u = \frac{\sum \overline{U}}{T_u} \quad P_d = \frac{\sum \overline{D}}{T_d} \quad E_u = \frac{\sum \overline{U}}{R_u} \quad E_d = \frac{\sum \overline{D}}{R_d}$$

where T_u and T_d are the total durations of the evaluation periods and R_u and R_d are the accumulated amounts of provisioned resources when scaling up and scaling down phases, respectively.[2] Table 2 shows the precision and efficiency of our framework.

Table 2. Basic operation efficiency

	Rate
P_u (Precision of scaling up)	99.2 %
P_d (Precision of scaling down)	99.1 %
E_u (Efficiency of scaling up)	99.6 %
E_d (Efficiency of scaling down)	99.4 %

Our component corresponds to an HTTP server, since web applications have very dynamic workloads generated by variable numbers of users, and they face sudden peaks in the case of unexpected events. Therefore, dynamic resource allocation is necessary not only to avoid application performance degradation but also to avoid under-utilized resources. The experimental results showed that our framework could follow the elastically provisioning and deprovisioning of resources quickly, and the number of the components followed the number of elastic provisioning and deprovisioning of resources exactly. The framework was scalable because its adaptation speed was independent of the number of servers.

6   Conclusion

This paper presented a framework for enabling distributed applications to be adapted to changes in their available resources in distributed systems. It was useful for adapting applications to elasticity in cloud computing. The key ideas behind the framework are *dynamic deployment of components* and *dividing and merging components*. The former enabled components to relocate themselves at new servers when provisioning the servers and at remaining servers when de-provisioning the servers, and the latter enables the states of components to be divided, and passed to other components, and merged with other components in accordance with user-defined functions. We believe that our framework is useful because it enables applications to be operated with elastic capabilities and resources in cloud computing.

References

1. N. Damianou, N. Dulay, E. Lupu, and M. Sloman: The Ponder Policy Specification Language, in Proceedings of Workshop on Policies for Distributed Systems and Networks (POLICY'95), pp.18–39, Springer-Verlag, 1995.
2. J. Dean and S. Ghemawat: MapReduce: simplified data processing on large clusters, in Proceedings of the 6th conference on Symposium on Opearting Systems Design and Implementation (OSDI'04), 2004.

---

[2] R_u and R_d correspond to the amount of provisioned resources according to cloud computing environments.

3. D. Garlan, S.W. Cheng, A.C.Huang, B. R. Schmerl, P. Steenkiste: Rainbow: Architecture-Based Self-Adaptation with Reusable Infrastructure, IEEE Computer Vol.37, No.10, pp.46-54, 2004.

4. B. Hindman, A. Konwinski, M. Zaharia, A. Ghodsi, A. Joseph, R. Katz, S. Shenker, and I. Stoica. Mesos: a platform for fine-grained resource sharing in the data center In Proceedings of USENIX Symposium on Networked Systems Design and Implementation (NSDI), 2011.

5. C. Inzinger, at al., Decisions, Models, and Monitoring–A Lifecycle Model for the Evolution of Service-Based Systems, In Proceedings of Enterprise Distributed Object Computing Conference (EDOC), pp.185-194, IEEE Computer Society, 2013.

6. M. A. Jaeger, H. Parzyjegla, G. Muhl, K. Herrmann: Self-organizing broker topologies for publish/subscribe systems, in Proceedings of ACM symposium on Applied Computing (SAC'2007), pp.543-550, ACM, 2007.

7. G. Jung, et. al.: A Cost-Sensitive Adaptation Engine for Server Consolidation of Multitier Applications, In Proceedings of Middleware'2009, LNCS, Vol.5896, pp.163183, Springer, 2009.

8. E. Lupu and M. Sloman: Conflicts in Policy-Based Distributed Systems Management, IEEE Transaction on Software Engineering, Vol.25, No.6, pp.852-869, 1999.

9. L. Lymberopoulos, E. Lupu, M. Sloman: An Adaptive Policy Based Management Framework for Differentiated Services Networks, in Proceedings of 3rd International Workshop on Policies for Distributed Systems and Networks (POLICY 2002), pp.147-158, IEEE Computer Society, 2002.

10. P. Mell, T. Grance: The NIST Definition of Cloud Computing, Technical report of U.S. National Institute of Standards and Technology (NIST), Special Publication 800-145, 2011.

11. A. Verma, L. Pedrosa, M. Korupolu, D. Oppenheimer, E. Tune, and J. Wilkes: Large-scale cluster management at Google with Borg, EuroSys15, ACM 2015.

12. U. Sharma, P. Shenoy, S. Sahu, A. Shaikh: A cost-aware elasticity provisioning system for the cloud In Proceedings of International Conference on Distributed Computing Systems (ICDCS'2011), pp.559570, IEEE Computer Society, 2011.

13. G. Tamura et. al.,: Towards Practical Runtime Verification and Validation of Self-Adaptive Software Systems, Proceedings of Self-Adaptive Systems, LNCS 7475, pp. 108132, 2013.

14. V. K. Vavilapalli, el. al.,: Apache Hadoop YARN: Yet Another Resource Negotiator, In Proceedings of Symposium on Cloud Computing (SoCC'2013), ACM, 2013.

15. World Wide Web Consortium (W3C): Composite Capability/Preference Profiles (CC/PP), http://www.w3.org/TR/NOTE-CCPP, 1999.

A Dynamic Model to enhance the Distributed Discovery of services in P2P Overlay Networks

Adel Boukhadra, Karima Benatchba, and Amar Balla

Ecole national Supérieur d'Informatique, BP 68M, Oued-Smar, 16309, Alger, Algérie
a_boukhadra@esi.dz, k_benatchba@esi.dz, a_balla@esi.dz

Abstract. In Service Computing (SC), online Semantic Web services (SWs) is evolving over time and the increasing number of SWs with the same function on the Internet, a great amount of candidate services emerge. So, efficiency and effectiveness has become a stern challenge for distributed discovery to tackle uniformed behavior evolution of service and maintain high efficiency for large-scale computing. The distributed discovery of SWs according to their functionality increases the capability of an application to fulfill their own goals. In this paper, we describe an efficient and an effective approach for improving the performance and effectiveness of distributed discovery of SWs in P2P systems. As most Web services lack a rich semantic description, we extend the distributed discovery process by exploiting collaborative ranking to estimate the similarity of a SWs being used by existing hybrid matching technique of OWL-S (Ontology Web Language for Services) process models in order to reduce costs and execution time. We mapped our distributed discovery of OWL-S process models by developing a real application based on Gamma Distribution; a technique used to decrease the bandwidth consumption and to enhance the scalability of P2P systems. The particularity of the Gamma Distribution is then integrated for disseminating request about the P2P networks to perform quality based ranking so that the best SWs can be recommended first. The experimental result indicates that our approach is efficient and able to reduce considerably the execution time and the number of message overhead, while preserving high levels of the distributed discovery of SWs on large-size P2P networks.

Keywords: SWs, Distributed Discovery, P2P Computing, Gamma Distribution, Matching of Ontology, OWL-S process model.

1 Introduction

Service-oriented computing (SOC) has emerged as an effective means of developing distributed applications and an accurate assessment the full potential of reputation which is essential for discovering between alternative Web services. Discovery of Web service is a great interest and is a fundamental area of research in distributed computing [6].

© Springer International Publishing AG 2017   13
C. Badica et al. (eds.), *Intelligent Distributed Computing X*,
Studies in Computational Intelligence 678, DOI 10.1007/978-3-319-48829-5_2

Actually, with the increasing number of SWs with the same function, in an open and dynamic environment, such as Internet, a great amount of candidate services emerge. Moreover, the number of registries that offer available Web services is also increasing significantly. It is not always easy to find services that matching users queries. Find a service from the candidate Web service set, according to the requirements of end-users has become a key hindrance and a strenuous task even for an experienced user. The Web service discovery satisfying the query is led by an optimization process, aiming to achieve the best SWs in the end [1] [4] [5] [7].

This could be very challenging to end-users given the huge number of available Web services online, who have to decide where to fulfill their requests: on local directory or to a replace this structure by several registries properly organized to support the dynamic, flexible and more efficient request propagation in a highly-dynamic distributed computing. Due to low accuracy, poor performance, high operational and maintenance cost, and sometimes for low availability of the functionality of SWs, the distributed discovery of SWs on the P2P systems provides a unique opportunity to address the above challenges [1] [7]. In recent years, the rise of P2P networks is attested by the increasing amount of interest in both commercial and academic areas to develop systems for data sharing simple and effective with many advantages such as decentralization, self-organization, autonomy, etc. At the same time, the SWs community has been slowly evolving toward a higher degree of distribution [4] [5] [6].

In this paper, we propose an efficient and an effective approach that addresses some aspects related to problems the time complexity of collaboration in the process of automatic discovery for SWs in P2P computing. For this purpose, our approach is based on P2P computing that proved to be scalable, more fault tolerant by eliminating the single point of failure, efficient by reducing the overhead of centralized update of the service discovery and robust solutions for distributed discovery of SWs. Specifically, our contributions in this work are summarized as follows. We exploit hybrid matching technique of OWL-S process models to develop functional features of Web services and the desired specification given by the user. Both the Web services and request can be then represented as OWL-S process model. We incorporate Gamma Distribution based collaborative ranking to identify additional functionally relevant Web services, in order to efficiently and effectively discover appropriate Web services distributed among all peers in a large-size P2P network. The idea of using Gamma Distribution is to manage large and continuously growing spaces of Web services with reasonable resolution times. The top-k most possible Web services are discovered, which are all considered as functionally relevant to the new Web services.

The remainder of this work is organized as follows: Section 2 describes the related work. In Section 3 we focus on the proposed distributed approach for discovering SWs in the unstructured P2P computing. An experimental evaluation is presented in Section 4, and, finally, in Section 5, we present our conclusion and highlight our future work directions.

2   Related Work

In this section, we discuss several representative related work and differentiate them with our work.

In [2], authors propose the P2P-based Semantic Driven Service Discovery (P2P-SDSD) framework to enable cooperation and communication based on a semantic overlay that organizes semantically the P2P-integrated knowledge space and emerges from local interactions between peers. The semantic overlay can be seen as a continuously evolving conceptual map across collaborative peers that provide similar services and constitute synergic service centers in a given domain. The semantic overlay enables effective similarity based service search and optimization strategies are defined for request propagation over the unstructured P2P network keeping low the generated network overload. Each collaborative peer in the unstructured P2P network has a local knowledge infrastructure constituted by: (i) UDDI Registry; (ii) Peer Ontological Knowledge, that provides a conceptualization of abstract service operations and Input/output parameters through a given domain ontology; a conceptualization of service categories through a Service Category Taxonomy (SCT).

In [7], present a technique to improve discovery in unstructured P2P service networks, based on a probabilistic forwarding algorithm driven by network knowledge, such as network density, and traces of already discovered service compositions (CONs). The technique aims at reducing the composition time and the messages exchanged during composition, relying on two considerations: if the network is dense, forwarding can be limited to a small number of neighbors; if the network is semi structured in CONs, forwarding can be directed to the super peers that may own the desired information. The approach improves the discovery and composition process by using distributed bidirectional search. The benefit is twofold: first, it is possible to have concurrent searches in a P2P service network in both goal directions (from pre- to post- and from post to preconditions), reducing the response time when solutions are present; second, when no complete solution for a goal is present, gaps in partial found solutions can be identified. This way, it is possible to have feedbacks about users' most required unavailable business operations, allowing providers to discover new business opportunities.

In [11], authors present a distributed approach to SWs publication and discovery by leveraging structured P2P network. In this work, the computers concerned constitute a P2P network to maintain the sharable domain and service ontologies to facilitate SWs discovery. When a requestor submits a semantic query for desired services, the P2P network can effectively obtain semantically qualified services. The main contributions of this work can be summarized as follows: this approach introduces a semantic-based service matching rule. In order to achieve the optimal match between a query, it proposes a concept of Ordered-Concept-Tree (OCT) to semantically sort the relevant concepts for service matching. In addition, to freely share and make full use of the semantic concepts defined in ontologies for OCT construction, it also proposes a method to publish ontologies to structured P2P network.

3   The Proposed Approach

In this section, we describe in detail the proposed approach that uses dynamic topology adaptation to improve the efficiency and the effectiveness of distributed discovery of SWs.

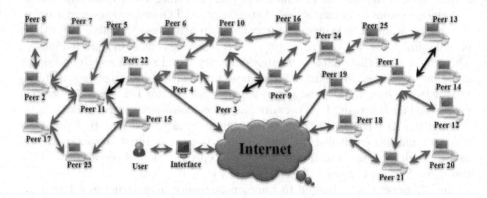

Fig. 1. Unstructured P2P Approach for Discovering SWs.

We propose further model which defines the topology of the P2P computing for supporting communication and collaboration between different Web service by the means of an effective use of the resources of P2P network and the management of scalability in an integrated and practical way. Such our model can support scalability, maximize search recall, reduce the query traffic and must also be able to achieve an acceptable performance. There are mainly three different architectures for P2P systems: hybrid, pure and unstructured; but the unstructured P2P systems are most commonly used in today's Internet [5] [6] [11].

In unstructured P2P network, we distinguish between two types of peers, neighbor and ultra-peer. Each peer is linked to a set of other peers in the P2P network via bi-directional connections, that is, each peer has a limited number of neighbors and it can choose the appropriate other neighbors it wishes to connect when necessary. When a peer receives a search request, it manages the discovery process locally and/or it will forward the query from other neighbors in the P2P network using the dynamic querying algorithm.

If a peer frequently returns good results of a large number of queries from one of its neighbors, it is likely to have common collaboration with the requesting peer to discover the appropriate Web services. It will mark this neighbor (the requesting peer) as an ultra-peer of the requesting peer (see Figure 1). To illustrate our model, consider the sample network depicted in Figure 1, in which each peer has a several neighbors and ultra-peers. Peer 11 has Peer 2, Peer 5, Peer 15 and Peer 17 as neighbors; and Peer 11 as ultra-peer.

In particular, our distributed discovery mechanism exploits metrics to broadcast queries in the P2P network, by associating a TTL[1] and $\Psi(\alpha, \lambda)$[2] value to the query messages; bigger values increase the success rate but may quickly lead to P2P network congestion. We then use this analysis to propose an algorithm that seeks to hit the minimum number of peers necessary to obtain the desired number of results for a given search. The general algorithm is described as follows.

Algorithm 1 : Distributed Discovery of SWs ();

1: **Input:** Upon reception of user request (R) at peer such as Search the SWs:(Input, Output, Precondition, Result and TextDescription).
2: **Output:** A set of SWs which responds to the user request.
3: **Begin**
4: **Discovery-Matching ();**
5: **if (There is a local SWs) then** // to discover a local SWs in this peer
6:     Send the Favorable Response;
7: **Else**
8: Calculate the TTL;
9: Calculate the $\Psi(\alpha, \lambda)$;
10: **if** (TTL ≥ 1) And ($\Psi(\alpha, \lambda) \geq Threshold$) **then**
11:     $TTL \leftarrow TTL - 1$;
12: **P2P-Discovery ();** // to propagate the query in ultra-peers or neighbors
13: **Else**
14: Drop the request;
15: **EndIf**
16: **EndIf**
17: **End**

Each peer in the unstructured P2P network executes the following main algorithm when receiving a request. Initially, each peer executes the main algorithm from step 1 to step 4 where it tries to response to the request locally (local basic SWs) and it sends the response to the transmitting peer (step 5 and step 6 of the main algorithm). If there is not a possibility to answer the request locally, the peer starts a distributed P2P discovery (step 12 of the main algorithm).

In addition, if the peer has no positive results, it decreases the TTL value by 1 and confirms $\Psi(\alpha, \lambda) \geq Threshold$, then it sends query towards all its immediate ultra-peers with the TTL value. The ultra-peer then executes the operations of TTL and $\Psi(\alpha, \lambda)$, to propagate the query to its other ultra-peers. If the ultra-peer has no positive results, then the ultra-peer forward the message contains the new TTL value to its ultra-peers. If all ultra-peers fail, the request will be forwarded to the next neighbors where the process is repeated until all neighbors are accessed or a positive result is given. If no positive result is found, the result sent by the last peer consultation will be negative.

---

[1] **TTL** : Time to Live.
[2] Gamma function.

One of the major issues in our approach is the ability to determine the degree of similarity between the concepts of two OWL-S process models. We investigate the use of hybrid matching component to increase the number of matching between the requests and OWL-S process model available in the OWL-S repository of different peers in the P2P network to improve system responses by the system that is asked. Each peer implements a number of SWs described semantically with OWL-S process model for publishing and discovering SWs in a local repository. The current trend of most matching systems is to combine the different matching techniques of ontologies OWL-S. The rationale behind the application of hybrid matching techniques in the scenario considered here is that we need two kinds of matching information to label semantic similarity links, in order to support efficient and effective SWs discovery.

We say that there is a similarity between concepts of Request R and Web service S if there is a degree of match between the set of concepts annotating the parameter (*Input, Output, Precondition, Result* and *TextDescription*) of Request R and the set of concepts annotating the same parameter (*Input, Output, Precondition, Result* and *TextDescription*) of Web service S.

To do so, we develop the procedure **Discovery-Matching()** (See algorithm 1) which creates a vector of semantic links in order to solve a service discovery request. The vector of semantic links stores the semantic similarity links established between the different SWs in a peer of the P2P network and the user request. In the remaining, we use the well-known similarity metric to compute the similarity assessment [10]:

$$Sim(R, S) = 1 - [2 \times \omega(lca(R, S)) - \omega(R) - \omega(S)] \tag{1}$$

where $lca(R, S)$ is the least common ancestor of R and S, and $\omega(R/S) \in [0, 1]$ is the weight of the concept R (or S) in the ontology. The similarity between two sets of concepts is then the average similarity of the mapping that maximizes the sum of similarities between the concepts of these sets. However, the similarity between $R = (In_R, Out_R, Pre_R, Res_R, Tex_R)$ and $S = (In_S, Out_S, Pre_S, Res_S, Tex_S)$ is the weighted average of their *Input, Output, Precondition, Result* and *TextDescription* similarities, as formalized by evaluating the following function [3]:

$$Similarity(R, S) = (\omega_{In} \times Sim(In_R, In_S) + \omega_{Out} \times Sim(Out_R, Out_S)$$
$$+\omega_{Pre} \times Sim(Pre_R, Pre_S) + \omega_{Res} \times Sim(Res_R, Res_S) + \omega_{Tex} \times Sim(Tex_R, Tex_S) \tag{2}$$

Where $Sim(R, S)$ is the function computing the similarity between two sets of parameters. Weights ($\omega_{In}, \omega_{Out}, \omega_{Pre}, \omega_{Res}$ and ω_{Tex}) designate, respectively the weight of *Input, Output, Precondition, Result* and *TextDescription* to evaluate the similarity ($\omega_{In} + \omega_{Out} + \omega_{Pre} + \omega_{Res} + \omega_{Tex} = 1$). This measure of similarity is probably the most widely used today and the most effective way to determine the semantic proximity between two concepts. This measure is used due to its high performances for process model matchmaking.

In our P2P approach, to find a resource, a peer broadcast a message to its ultra-peers or neighbors. However, If all remote peers sends the query by flooding, than a serious problem of flooding emerges due to the excessive traffic overheads caused by a large number of redundant message forwarding, increasing the bandwidth consumption and reducing the performance of the P2P system; particularly in a P2P system with a high connectivity topology. Although it is effective in our P2P approach, flooding is very inefficient because it results in a great amount of redundant messages. To optimize the flooding, some criteria have to be defined to select the peer for which the request will be sent. Each peer must limit the number of simultaneous transfers of request.

Typically two connected peers each ensures the other to be active in the P2P network and begins sending requests with the maximum interval of time defined by the session time. The session time of peer could be modeled by an exponential distribution. In this work, our aim is to achieve an extremely high network scalability and enhancing the other network performance metrics. For this purpose, we analyse how this widespread property of P2P networks can be more accurately described by Gamma Distribution that dynamically adapts the network effective density to enable a large scale deployment of the P2P network while providing good overall performances.

From the function computing the similarity between two sets of concepts $Similarity(R, S) \in [0, 1]$ and given peer TTL value, we have defined the following evaluating Gamma Distribution function as follows:

$$P(session < t) = \Psi(t|\alpha, \lambda) = \frac{\lambda^\alpha}{\Gamma(\alpha)} \times \exp^{-(\frac{t}{\lambda})} \qquad (3)$$

In a P2P network of N peers with D being the average peer degree, we experiment with several values of the TTL to describe the shape α and the scale λ parameters when:

$$\Gamma(\alpha) = \int_o^\infty x^{(\alpha-1)} \exp^{-x} dx; \alpha = \frac{\ln(N-1)}{D^{TTL}}; \lambda = \frac{1}{TTL+1} \qquad (4)$$

We opt for a Gamma function that uses exponential function $(\exp^{-(\frac{t}{\lambda})})$ because a high Gamma function incurs more redundant rebroadcast while a low Gamma function leads to low reachability. Moreover, peers with low values of TTL, α and λ should be assigned a high Gamma function while those with high values of TTL, α and λ are assigned a low Gamma function. Therefore, as the number of neighbors increases, the Gamma function should decreases.

Our motivation for this Gamma function is to enhance rebroadcast decision by taking into account key network parameters and peer information through TTL, α and λ value. We implement our matching technique for Gamma function that can be used to assess the similarity function $(Similarity(R, S) \in [0, 1])$ between two concepts of OWL-S process models and to provide an alignment between them; 0 means the concepts are totally different, 1 means that they are totally similar. Consequently, the similarity between concepts of Request R and Web service S is defined on the basis of their semantic relationship in the ontology.

We apply results achieved by statistical analysis of random graphs to selecting appropriate initial value for TTL of the query. Applying the results provided by [9], each peer decides on the appropriate TTL value for its queries based on the information collected locally. The TTL between two randomly chosen peers on any P2P network is approximated as follows:

$$\mathbf{TTL} = \frac{\ln[(N-1)(Z_2 - Z_1) + Z_1^2] - \ln(Z_1)^2}{\ln(Z_2/Z_1)} \tag{5}$$

Where Z_i is the number of neighbors which are i hops away from the originator peer. TTL between two peers presents a reasonable estimation of the distance between the originator of the query and the peer that eventually serves the requested object. TTL is actually the scope of the request: more important it is, the more there will be peers that will be visited, and the request will likely be satisfied. A large TTL also causes average response greater. Using a large TTL, this type of infrastructure can meet a maximum of elements corresponding to the search criteria. In addition, this approach is fast and reliable (it behaves very well in highly dynamic networks with many arrivals or departures of peers). In the next section we will proved our approach, evaluation results and share our experiences.

4 Experiments and Results

We conducted a series of experiments to demonstrate and evaluate the effectiveness and the efficiency of our scalable approach for discovering SWs in the unstructured P2P network through event-driven simulations, which are usually used to evaluate the performance of large-scale P2P systems. For this reason, we are using PeerSim simulator [8] to simulate an unstructured P2P network. The experimentation has been performed with the number of peers that varies in the range from 100 to 1500 with an iteration range of 200. By generated requests we mean the total number of overall requests produced and forwarded on the unstructured P2P network as a consequence of a request submitted to a peer; this parameter depends on the number of peer in the network and on the peers average number of connections to its neighbors. We run our experiments on computers with Intel Core i5 CPU (2.54 GHz and 4GB RAM) under Windows 7. The data used in our experiments are OWLS-TC[3] 4.0. It provides 1083 SWs written in OWL-S 1.1. It provides a set of 42 test queries which are associated with relevance sets to conduct performance evaluation experiments. With regard to the different similarity measures which are implemented in our approach, we used the Java API SIMPAC[4] (Similarity Package) which represents a comprehensive library that contains all the important similarity measures. We use JWordNetSim to measure the similarity between synsets in WordNet[5] 2.0.

---

[3] http://www.semwebcentral.org/projects/owls-tc/

[4] http://sourceforge.net/projects/simmetrics

[5] http://wordnet.princeton.edu/

To evaluate the efficiency and scalability of our approach, we must compare our approach with a simple flooding protocol Gnutella P2P protocol [1]. In our experiment, we study the significance of our algorithm in terms of computation time. We define the optimality ratio as follows :

$$\textbf{Optimality Ratio} = \frac{W_{Gamma} - W}{W_{Gamma}} \qquad (6)$$

where W is the execution time of our algorithm without Gamma function and W_{Gamma} is the execution time of our algorithm by applying Gamma function. The execution time has been measured upon the number of invokes that our approach, reflecting the number of discovering SWs in unstructured P2P network. This helps demonstrate the scalability of our approach. The execution time has been measured using The Eclipse Test and Performance Tools Platform[6]. Experimentation results will be analyzed in the following.

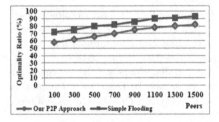 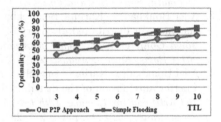

Fig. 2. Optimality Ratio vs the Peer.

Fig. 3. Optimality Ratio vs the TTL.

The graph in Figure 2 or 3 illustrates some of the performance test results. We notice that the optimality ratio increases slightly along with the number of peers or TTL. The results reported in these figures represent the average computation times found for each given request. Despite the exponential theoretical complexity, the figure 2 or 3 shows that our scalable algorithm can be used, with acceptable computation time. More clearly shown in Figure 2 or 3, Gnutella protocol has a very high computation time because it uses simple flooding algorithm. The simple flooding algorithm leads to high computational time. Our P2P approach performs better than the Gnutella protocol in terms of optimality ratio because it uses Gamma function that reduces considerably the computation time.

---

[6] TPTP: http://www.eclipse.org/tptp/

5   Conclusion and Future Work

In this paper, we have discussed efficient and effective approach, for improving the performance of distributed and cooperative discovery of SWs in the unstructured P2P networks. We have proposed the matching technique of OWL-S to fulfill the users requirements and the Gamma Distribution which reduces the query traffic. In the future work, we will focus our efforts on optimizing the communicational complexity by using optimization techniques such as heuristics and meta heuristics.

References

1. Hassan Barjini, Mohamed Othman, Hamidah Ibrahim, and Nur Izura Udzir. Hybridflood: minimizing the effects of redundant messages and maximizing search efficiency of unstructured peer-to-peer networks. *Cluster Computing*, 17(2):551–568, 2014.
2. Devis Bianchini, Valeria De Antonellis, and Michele Melchiori. P2p-sdsd: on-the-fly service-based collaboration in distributed systems. *International Journal of Metadata, Semantics and Ontologies*, 5(3):222–237, 2010.
3. Adel Boukhadra, Karima Benatchba, and Amar Balla. Hybrid ontology-based matching for distributed discovery of sws in p2p systems. In *16th IEEE International Conference on High Performance Computing and Communication, HPCC 2014, Paris, France, August 20-22, 2014*, pages 116–123. IEEE, 2014.
4. Adel Boukhadra, Karima Benatchba, and Amar Balla. Efficient P2P approach for a better automation of the distributed discovery of sws. In *Distributed Computing and Artificial Intelligence, 12th International Conference, DCAI 2015, Salamanca, Spain, June 3-5, 2015*, pages 277–284, 2015.
5. Adel Boukhadra, Karima Benatchba, and Amar Balla. Efficient distributed discovery and composition of OWL-S process model in P2P systems. *J. Ambient Intelligence and Humanized Computing*, 7(2):187–203, 2016.
6. P Fragopoulou, C Mastroianni, R Montero, A Andrjezak, and D Kondo. Self-* and adaptive mechanisms for large scale distributed systems. In *Grids, P2P and Services Computing*, pages 147–156. Springer, 2010.
7. Angelo Furno and Eugenio Zimeo. Efficient cooperative discovery of service compositions in unstructured p2p networks. In *Parallel, Distributed and Network-Based Processing (PDP), 2013 21st Euromicro International Conference on*, pages 58–67. IEEE, 2013.
8. Alberto Montresor and Márk Jelasity. Peersim: A scalable p2p simulator. In *Peer-to-Peer Computing, 2009. P2P'09. IEEE Ninth International Conference on*, pages 99–100. IEEE, 2009.
9. Mark EJ Newman, Steven H Strogatz, and Duncan J Watts. Random graphs with arbitrary degree distributions and their applications. *Physical review E*, 64(2):026118, 2001.
10. Pavel Shvaiko and Jérôme Euzenat. Ontology matching: state of the art and future challenges. *Knowledge and Data Engineering, IEEE Transactions on*, 25(1):158–176, 2013.
11. Huayou Si, Zhong Chen, Yong Deng, and Lian Yu. Semantic web services publication and oct-based discovery in structured p2p network. *Service Oriented Computing and Applications*, 7(3):169–180, 2013.

Simulation of Dynamic Systems Using BDI Agents: Initial Steps

Amelia Bădică[1], Costin Bădică[1], Marius Brezovan[1], and Mirjana Ivanović[2]

[1] University of Craiova, A.I.Cuza 13, 200530, Craiova, Romania
ameliabd@yahoo.com, {cbadica,mbrezovan}@software.ucv.ro
[2] University of Novi Sad, Faculty of Sciences, Novi Sad, Serbia
mira@dmi.uns.ac.rs

Abstract. In this paper we propose a framework based on BDI software agents for the modeling and simulation of dynamic systems. The target system is broken down into a number of interacting components. Each component is then mapped to a BDI agent that captures its behavioral aspects. The system model is described as a multi-agent program that is specified using the Jason agent-oriented programming language.

Keywords: dynamic system simulation, BDI agent, agent-oriented programming

1 Introduction

Multi-agent systems were proposed as a new paradigm for the modelling, simulation and programming of complex systems. Software agents proved appropriate for the engineering of complex contemporary applications that are composed of many, possibly heterogeneous, interacting, fault tolerant and distributed components, that operate in highly dynamic and uncertain environments [7]. During the last decade researchers were interested in investigating the relationships between multi-agent systems and dynamic systems regarding modeling, simulation and implementation aspects [3, 9, 5].

Generally by dynamic system we understand a system that contains elements that can change in time. Dynamic systems have many applications in engineering, biology, economy and sociology. In this paper we focus on dynamic systems that are characterized by a state-space representation as a set of differential equations. The goal of our research is to propose a framework based on BDI software agents [10] for the modeling and simulation of dynamic systems.

The target dynamic system is broken down into a number of interacting components that represent the functional blocks of the system model. Each component has a given type that precisely characterizes its behavior as a (possibly state-based) mathematical function. Consequently, a component has a finite number of inputs and one output.

Each component is then mapped to a BDI agent that captures its behavioral aspects. The system model is described as a multi-agent program that is specified using the Jason agent-oriented programming language [2]. Agents exchange simulation information following a data-flow paradigm, initially proposed in [4]. The multi-agent system can be used to perform the distributed simulation of the target dynamic system on a computer network.

© Springer International Publishing AG 2017   23
C. Badica et al. (eds.), *Intelligent Distributed Computing X*,
Studies in Computational Intelligence 678, DOI 10.1007/978-3-319-48829-5_3

Each agent encapsulates information and knowledge to support its functionality. Agents are configured to include the following information: the agent type, the set of parameters completely defining the mathematical function performed by the agent, the component state, as well as the agent acquaintance model that is necessary for agent interaction. The agent knowledge includes a set of plans that supports the agent to achieve its mathematical function. The information that is dynamically generated during the simulation process is exchanged by agents via messages and it is temporarily stored by the belief base of each agent.

The proposed framework can be used for the simulation, as well as for the implementation of dynamic control systems [7] using agent-based state-of-the-art software technologies: distributed multi-agent platforms and agent-oriented programming languages [1].

2 Model and Implementation

2.1 Dynamic Systems

In this paper we are interested in dynamic systems that are characterized by a state-based representation using first order differential equations (1).

$$\dot{x}(t) = f(x(t), u(t)) \tag{1}$$

Here $x(t)$ denotes that state vector of the system and $u(t)$ denotes the system input. By $\dot{x}(t)$ we denotes the first derivative of x. Usually $x : [t_0, \infty) \rightarrow \mathbb{R}^n$ and $u : [t_0, \infty) \rightarrow \mathbb{R}^m$. Here n represents the number of state variables and m represents the number of system inputs. $f : \mathbb{R}^n \times \mathbb{R}^m \rightarrow \mathbb{R}^n$ is known as the state transition function of the system. If f is continuously differentiable and u is piecewise continuous then there is a unique solution of equation (1) that satisfies the initial condition $x(t_0) = x_0$ [8]. Sometimes variable t is omitted in equation (1), thus the equation being simplified as $\dot{x} = f(x, u)$. In what follows we are going to follow this writing convention.

Let us consider for example the model of an ecological system containing only two species that behave according to a predator-prey relationship, as described in [8]. Let us denote with $x_i(t)$ the number of individuals of each species at time point t for $1 \leq i \leq 2$. Let $\lambda_i > 0$ and $\mu_i > 0$ be the birth rates and respectively the mortality rates of the two species. Therefore $\lambda_i x_i(t)$ individuals are born and respectively $\mu_i x_i(t)$ are dying by being eaten per time unit for each species $1 \leq i \leq 2$.

The mortality rates μ_i of each species generally depend on the availability of food and the risk of being eaten, which in turn depend on the number of individuals of each species, so:

$$\mu_i = \mu_i(x_1, x_2) \tag{2}$$

Combining equations (1) and (2) we obtain equation (3) that describes our ecological system:

$$\begin{aligned} \dot{x}_1 &= (\lambda_1 - \mu_1(x_1, x_2))x_1 \\ \dot{x}_2 &= (\lambda_2 - \mu_2(x_1, x_2))x_2 \end{aligned} \tag{3}$$

Now, following [8], in a predator-prey scenario let assume that species 1 preys species 2. Then the mortality rate of the predator (species 1) decreases when the number of individuals of the prey (species 2) increases. Moreover, the mortality rate of the prey (species 2) increases when when the number of individuals of the predator (species 2) increases. These aspects can be captured using equations (4) ($\alpha_1 < 0$ and $\alpha_2 > 0$.

$$\mu_1(x_1, x_2) = \gamma_1 + \alpha_1 x_2$$
$$\mu_2(x_1, x_2) = \gamma_2 + \alpha_2 x_1$$
(4)

Substituting equations (4) in (2) we obtain the model of the predator-prey dynamic system described by (5).

$$\dot{x}_1 = (\lambda_1 - (\gamma_1 + \alpha_1 x_2))x_1$$
$$\dot{x}_2 = (\lambda_2 - (\gamma_2 + \alpha_2 x_1))x_2$$
(5)

2.2 Block Diagrams

The mathematical analysis of dynamic systems can be achieved either using analytic solutions or using simulation. Very often an analytic solution is not available so the only feasible approach that remains is simulation.

Following [8], we can simulate equation (1) by solving the integral equation (6):

$$x(t) = \int_{t_0}^{t} f(x(\tau), u(\tau))d\tau$$
(6)

Equation (6) contains two elements: i) the algebraic block f, and ii) the integration block. So the equation can be represented as a simple block diagram, as shown in Figure 1. Note that arrows on this diagram represent vectorial flows, as follows: x and $\dot{x}$ of size n and respectively u of size m.

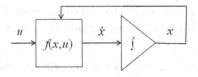

Fig. 1. General block diagram of a dynamic system.

The algebraic block f is usually decomposed into a set of interconnected elementary blocks that achieve various mathematical functions including: summation, multiplication, constant input, nonlinear elements, etc. Some examples of elementary blocks (including the integration block) are presented in Figure 2. Note that a block can be either parameterized or not parameterized. For example blocks shown in Figures 2a and 2d are parameterized, while blocks shown in Figures 2b and 2c are not parameterized.

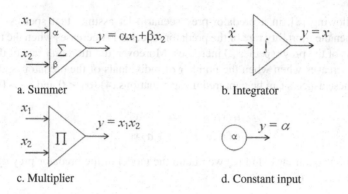

Fig. 2. Examples of elementary blocks.

2.3   AgentSpeak and Jason Multi-Agent Systems

AgentSpeak(L) is an abstract agent-oriented programming language initially introduced in [10]. Jason is a Java-based implementation, as well as an extension of AgentSpeak(L) [2].

AgentSpeak(L) follows the paradigm of practical reasoning, i.e. reasoning directed towards actions, and it provides an implementation of the belief-desire-intention (BDI) architecture of software agents.

According to this view an agent is a software module that (i) provides a software interface with the external world and (ii) contains three components: belief base, plan library and reasoning engine.

The agent's external world consists of the physical environment, as well as possibly other agents. The set of agents known by a given agent at a given time point defines the agent's acquaintance model. Consequently, the agent interface provides three elements: sensing interface, actuation interface and communication interface. The agent uses its sensing interface to get percepts from its physical environment. The agent uses its actuation interface to perform actions on its physical environment. Finally, the agent uses its communication interface to interact by exchanging messages with other known agents.

Belief base It defines what an agent "knows" or "believes" about its environment at a certain time point. The BDI architecture does not impose a specific structuring of the belief base other than as a generic container of beliefs.

By default Jason uses a logical model of beliefs by structuring the belief base as a logic program composed of facts and rules. An atomic formula has the form $p(t_1, \ldots, t_n)$ such that p is a predicate symbol of arity $n \geq 0$, and t_i are logical terms for $i = 1, \ldots, n$. A belief is either a fact represented by an atomic formula a or a rule $h : -b_1 \& \ldots \& b_k$, $k \geq 1$ such that h and b_j, $j = 1, \ldots, k$ are atomic formulas.

Plan library It defines the agent's "know-how" structured as a set of behavioral elements called plans. A plan follows the general pattern of event-condition-action rules and it is composed of three elements: triggering event, context and body. A plan has the form $e : c < - b$ where e is the triggering event, c is a plan context and b is a plan body.

A *plan body* specifies a sequence of agent activities $a_1 ; \ldots ; a_m$. Each a_i, $1 \le i \le m$, designates an agent activity. AgentSpeak(L) provides three types of activities: actions, goals, and belief updates. Actions define primitive tasks performed by the agent either on the environment (external actions) or internally (internal actions). Goals represent complex tasks. AgentSpeak(L) distinguishes between test goals, represented as $?a$ and achievement goals, represented as $!a$, where a is an atomic formula. Belief updates represent the assertion $+b$ or the retraction $-b$ of a belief b from the belief base.

A *plan context* is represented by a conjunction of conditions, such that each condition is either an atomic formula a or a negated atomic formula not a. The operator not is interpreted as "negation as failure" according to the standard semantics of logic programming.

A *triggering event* specifies a situation that can trigger the selection of a plan execution. An event can represent i) an assertion $+b$ or a retraction $-b$ of a belief b from the belief base or ii) an adoption $+g$ or dropping $-g$ of a goal g.

A plan will be actually selected for execution if and only if its context logically follows from the belief base. If $\mathcal{B}$ is the current belief base then this condition can be formally expressed as $\mathcal{B} \models c$.

Reasoning engine Each Jason agent contains a "reasoning engine" or "agent interpreter" component that controls the agent execution by "interpreting" the Jason code. The reasoning engine performs a reasoning cycle consisting of a sequence of steps: perceives the environment, updates its belief base, receives communication from other agents, selects an event, selects an applicable plan and adds it to its agenda, selects an item (intention) for execution from the agenda, and finally executes the next action of the partially instantiated plan that represents the top of the currently selected intention.

The agent agenda is structured as a list of intentions. We can think of each intention as a stack of partially instantiated plans (somehow similar to a call stack in imperative programming) that represents an agent execution thread. So each stack represents one focus of attention of the agent. Using this approach an agent can execute concurrent activities by managing multiple focuses of attention [2].

Note that a partially instantiated plan that was created by invoking an achievement goal with $!g$ is stacked on top of the intention that invoked it. This means that this plan will be executed in the focus of attention corresponding to the invoking intention. Alternatively, an invocation with $!!g$ will create a new focus of attention for it, thus increasing the number of tasks that are executed concurrently by the agent.

2.4 Mapping Blocks to Agents

In the early days, simulation was realized using analog computers. They allowed the natural mapping of elementary blocks onto analog electronic blocks inside the analog computer, thus enabling to exploit the natural parallelism that is intrinsically present in a dynamic system. Basically data flows were mapped to electronic signals, while system simulation was achieved in terms of purely physical terms that quantify analog electronic signals.

Digital simulation provides methods for mapping a dynamic system model to simulation software that is usually developed using general purpose programming languages enhanced with numerical processing libraries. In particular, this method heavily employs software packages for numerical integration based on traditional algorithms, for example Euler or Runge-Kutta families of numerical methods [6].

Multi-agent systems are useful for modeling systems of interconnected components by providing a natural mapping of system components to a set of interacting software agents. Here we propose a new modelling paradigm of dynamic systems inspired by the rule-based approach initially proposed in [4], that combines the characteristics of both analog and digital simulation into a single unified approach.

Each elementary system block of a dynamic system model is mapped to a Jason agent. The agent incorporates the following elements:

i) A belief base that contains:
   a) A set of facts that define the parameters (if any) of the block;
   b) A fact that defines the type and the name of the block;
   c) A fact that defines the agent's acquaintance model; we assume that the acquaintance model is static, i.e. once defined, it will not change during the simulation;
   d) A set of temporary facts that define the internal state of the block. This information depends on the block type and it supports the block to carry out its function. If present, it is usually updated during the simulation.
ii) A set of template plans that support the agent to achieve its mathematical function. Basically, a template plan is responsible with processing the incoming information received by the agent, according to the agent's function. Then, the resulted output information is dispatched to the agent's acquaintances.

Each block is introduced using the predicate block(Type, BlockIndex, Inputs) that defines the block type, the block name (the same as the agent name that corresponds to the block), as well as the list of block inputs. By default each block has a single output, so there is no need to specify it in the block definition.

The information exchanged by the agents is represented using the predicate message(Iteration, Time, BlockIndex, Input, Value). Whenever a new value is produced by an agent, that value is packed into a message containing the time point, the destination block name, the corresponding input of the destination block, and the value itself. Then the message is dispatched to the destination agent.

A multiplier agent, as introduced in Figure 2c, contains the following template plan:

```
@multiplier[atomic] +!advance :
  block(multiplier,BlockIndex,[Input1,Input2]) &
  message(I,Time,BlockIndex,Input1,Value1) &
  message(I,Time,BlockIndex,Input2,Value2)
  <-
  -message(I,Time,BlockIndex,Input1,Value1)[source(_)];
  -message(I,Time,BlockIndex,Input2,Value2)[source(_)];
  ?destination(Ds);
  !dispatch(I,Time,Value1*Value2,Ds);
  !!advance.
```

Parameters of parameterized blocks are usually attached to block inputs. For example, the summer takes two parameters α and β, as shown in Figure 2a. The parameters are defined using the predicate `parameter(BlockIndex, Input, ParamValue)` that introduces the parameter value, as well as the block input to which the parameter is assigned.

A summer agent contains a template plan defined as follows:

```
@summer[atomic] +!advance :
   block(summer,BlockIndex,[Input1,Input2]) &
   message(I,Time,BlockIndex,Input1,Value1) &
   message(I,Time,BlockIndex,Input2,Value2)
   <-
   ?parameter(BlockIndex,Input1,Param1);
   ?parameter(BlockIndex,Input2,Param2);
   -message(I,Time,BlockIndex,Input1,Value1)[source(_)];
   -message(I,Time,BlockIndex,Input2,Value2)[source(_)];
   ?destination(Ds);
   !dispatch(I,Time,Value1*Param1+Value2*Param2,Ds);
   !!advance.
```

Note that purely algebraic blocks do not advance the simulation time, i.e. their processing is carried out within the current simulation iteration. Normally, in digital simulation, the integration part is responsible with advancing the simulation time. With our approach, the integration is "distributed" to the team of integrator agents. Basically these agents have to decide the simulation time step, according to the integration algorithm. This time step defines the current simulation time point that is used by the input agents, i.e. those agents that are responsible with generating the inputs of the dynamic system according to function u from Figure 1. The input agents must know the value of the current time point in order to generate the inputs of the dynamic system.

Note that a dynamic system might have more inputs, so an agent-based simulation model might have more input agents. In order to avoid the redundant replication of the time management function in each input agent, we created a unique separate timer agent that is responsible with time management. This agent "talks" with the team of integrator agents to get the simulation time step, and it then "spreads" this value to the input agents.

Note that while Euler's integration method uses a very simple equation for updating the values of the derivatives of f (see Figure 1) based on a fixed simulation time step $h > 0$, other integration methods are using more complex approaches. For example, the Runge-Kutta family of methods uses several sub-stages within the same iteration; in each substage a fixed step ch is used with a series of specific $c < 1$ values that are defined for each type of Runge-Kutta method. Moreover, there are integration algorithms that adjust adaptively the integration step in order to meet a certain integration error. So, the problem of coordinating the integrator agents and the timer agent can be more complicated in the general case.

Nevertheless, to keep things simple, in what follows we assume that our integrator agents use Euler's method that is based on equation (7). The first integrator that advances the simulation time will notify the timer agent. The time agent will subsequently inform the other input agents, and a new simulation iteration is thus triggered.

$$x(t + h) = x(t) + hf(x(t), u(t)) \tag{7}$$

Differently from purely algebraic blocks, an integrator agent must define additional facts in its belief base: i) `step(Step)` to represent the simulation time step; ii) `state(BlockIndex, CurrentTime, CurrentValue)` to represent the value of its current state; this is a temporary fact that must be initialized with the initial time of the simulation and with the value of the initial state of this integrator; iii) `endtime(EnTime` to define the time when the simulation ends.

An integrator agent that uses Euler's method contains a template plan as follows:

```
@integrator[atomic] +!advance :
  block(integrator,BlockIndex,[Input]) &
  message(I,Time,BlockIndex,Input,DerivValue) &
  state(BlockIndex,Time,OldValue) &
  step(StepValue) &
  endtime(FinalTime) &
  Time+StepValue <= FinalTime
  <-
  -message(I,Time,BlockIndex,Input,DerivValue)[source(_)];
  NewTime = Time+StepValue;
  NewValue =  StepValue*DerivValue+OldValue;
  -state(BlockIndex,Time,OldValue);
  +state(BlockIndex,NewTime,NewValue);
  ?destination(Ds);
  !dispatch(I+1,NewTime,NewValue,Ds);
  !!advance.
```

In particular integrator agents are responsible with triggering the simulation start. An integrator agent specifies an initial achievement goal `!start` as follows:

```
+!start : true <-
  ?destination(Ds);
  ?block(integrator,BlockIndex,_);
  ?state(BlockIndex,Time,Value);
  !dispatch(1,Time,Value,Ds);
  !advance.
```

The timer agent manages a counter of iterations. This agent receives messages from the integrators. Each message contains new values of the iteration counter and the simulation time. Whenever a message contains a value of the iteration counter that is higher than its current value, the timer agent sends an update containing the new value of the simulation time to the input agents. Otherwise, the timer agent simply ignores this message. Its template plan is defined as follows:

```
@timer[atomic] +!advance :
  block(timer,BlockIndex,[Input]) &
  message(I,Time,BlockIndex,Input,_) &
  currentiter(ICrt) &
  I > ICrt
  <-
  -message(I,Time,BlockIndex,Input,_)[source(_)];
  -currentiter(ICrt);
  +currentiter(I);
  ?destination(Ds);
  !dispatch(I,Time,0.0,Ds);
  !!advance.
```

Finally, the constant input agent outputs a constant value that is defined using predicate `value(Constant)`. The template plan of this agent is shown below. Note that this agent only uses the time component of the received message, while the received value is simply ignored.

```
@input[atomic] +!advance :
  block(input,BlockIndex,[Input]) &
  message(I,Time,BlockIndex,Input,_)
  <-
  -message(I,Time,BlockIndex,Input,_)[source(_)];
  ?value(Constant);
  ?destination(Ds);
  !dispatch(I,Time,Constant,Ds);
  !!advance.
```

The plan for achieving the !dispatch goal is shared by all the agents and it is defined as follows:

```
+!dispatch(_,_,V,[]) : true.
+!dispatch(J,T,V,[[D,I] | Ds]) : true <-
  .send(D,tell,message(J,T,D,I,V));
  !dispatch(J,T,V,Ds).
```

3   Experiment and Discussion

In this section we present an experiment involving a multi-agent system that simulates the predator-prey model that was introduced in Section 2. Figure 3 introduces the block diagram of our multi-agent system.

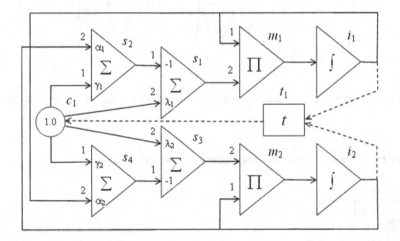

Fig. 3. Block diagram of the predator-prey dynamic system.

Our experimental multi-agent system contains 12 agents, as follows: 2 integrators i_1 (predator population state) and i_2 (prey population state); 4 summers s_1, s_2, s_3, and s_4; 2 multipliers m_1 and m_2; 1 timer t_1 (the agent represented using the box labelled with t); 1 constant input c_1; 2 printer agents p_1 and p_2, not shown in the figure. Each of the printer agents is connected to the output of the corresponding integrator. Printer agents are responsible with logging the system outputs represented by the outputs of the integrators in this case. Note that the continuous lines indicate the flows of simulation information, while the dashed lines indicate the flows of temporal information. Note also that agents with more than one input have their inputs uniquely identified by integer indexes, while agents with a single input have the index 1 assigned by default (not shown in the figure) to their input.

The parameters of our sample dynamic system model were chosen as follows: $\alpha_1 = -1.0, \gamma_1 = 2.0, \lambda_1 = 1.0, \alpha_2 = 1.0, \gamma_2 = 1.0, \lambda_2 = 3.0$. The parameters of the simulation experiment were chosen as follows: integration step $h = 0.001$, simulation time $T = 30.0$, initial time $t_0 = 0.0$, initial size of predator population $x_1(0) = 2$, and initial size of prey population $x_2(0) = 4$ (in thousands of individuals).

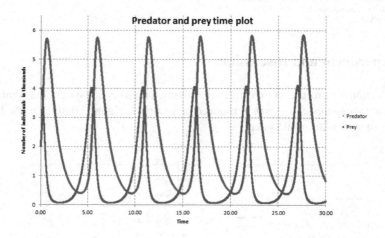

Fig. 4. Simulation results presenting the number of individuals of predator and prey populations.

The belief bases of some of these agents are shown below:

```
// Integrator i1
block(integrator,i1,[1]).
parameter(i1,1,1.0).
step(0.001).
state(i1,0.0,2.0).
destination([[m1,1],[s4,2],[p1,1],[t1,1]]).
endtime(30.0).
// Summer s1
block(summer,s1,[1,2]).
parameter(s1,1,-1.0).
parameter(s1,2,1.0).
destination([[m1,2]]).
// Constant input c1
block(input,c1,[1]).
destination([[s1,2],[s2,1],[s3,2],[s4,1]]).
value(1.0).
// Timer t1
block(timer,t1,[1]).
destination([[c1,1]]).
currentiter(0).
```

4 Conclusion

In this paper we proposed a new multi-agent approach based on BDI agents and Jason agent programming language for the modeling and simulation of dynamic systems. In our opinion this approach combines the strengths of both approaches of analog and digital simulation by providing a more natural mapping of the block diagram of a dynamic system to a distributed computational system. Our proposed mapping captures better than traditional simulation approaches the intrinsically parallelism that is present in the target dynamic system. In the future we plan to expand our approach to support more advanced numerical integration methods, as well as more types of blocks for describing more complex models of dynamic systems.

References

1. Costin Bădică, Zoran Budimac, Hans-Dieter Burkhard, Mirjana Ivanović: Software agents: Languages, tools, platforms. Comput. Sci. Inf. Syst. 8 (2): 255–298 (2011) doi: 10.2298/CSIS110214013B
2. Rafael H. Bordini, Jomi Fred Hübner, Michael Wooldridge: Programming Multi-Agent Systems in AgentSpeak using Jason. ser. Wiley Series in Agent Technology. John Wiley & Sons Ltd (2007)
3. Andrei Borshchev and Alexei Filippov: From System Dynamics and Discrete Event to Practical Agent Based Modeling: Reasons, Techniques, Tools. The 22nd International Conference of the System Dynamics Society, July 25 - 29, Oxford, England (2004)
4. Luc Boullart: A Rule-Based Simulator for Dynamic Systems. In: Artificial Intelligence Handbook: Applications. Volume 2 of Artificial Intelligence Handbook, 283-331, Instrument Society of America (1989)
5. Giancarlo Fortino, Francesco Rango, Wilma Russo, Corrado Santoro: Translation of statechart agents into a {BDI} framework for {MAS} engineering. Engineering Applications of Artificial Intelligence, 41, 287–297, Elsevier (2015) doi: 10.1016/j.engappai.2015.01.012
6. Richard W. Hamming: Numerical Methods for Scientists and Engineers. Dover Publications (1987)
7. Nicholas R. Jennings, Stefan Bussmann: Agent-based control systems: Why are they suited to engineering complex systems?. IEEE Control Systems, 23 (3), 61–73, IEEE (2003) doi: http://dx.doi.org/10.1109/MCS.2003.1200249
8. Lennart Ljung, Torkel Glad: Modeling of Dynamic Systems. ser. Prentice Hall Information and System Science Series. Prentice Hall (1994)
9. Charles M. Macal: To agent-based simulation from System Dynamics. In: Proceedings of the 2010 Winter Simulation Conference (WSC), 371–382, IEEE 2010 doi: 10.1109/WSC.2010.5679148
10. Anand S. Rao: AgentSpeak(L): BDI agents speak out in a logical computable language. In: Van de Velde, Walter and Perram, J. W. (eds.): Agents Breaking Away. Lecture Notes in Computer Science 1038, 42–55, Springer Berlin Heidelberg (1996) doi: 10.1007/BFb0031845
11. Ilkka Seilonen, Teppo Pirttioja, Kari O. Koskinen: Extending process automation systems with multi-agent techniques. Engineering Applications of Artificial Intelligence, 22 (7), 1056–1067, Elsevier (2009) doi: 10.1016/j.engappai.2008.10.007

Part II

Internet of Things

A Multi-Agent Approach for the Deployment of Distributed Applications in Smart Environments

Ferdinand Piette[1,2], Costin Caval[1], Cédric Dinont[2], Amal El Fallah Seghrouchni[1], and Patrick Taillibert[1]

[1] Sorbonne Universités, UPMC Univ Paris 06, LIP6, Paris, France,
[2] Institut Supérieur de l'Électronique et du Numérique, Lille, France

Abstract. This paper presents an approach for the configuration, deployment and monitoring of distributed applications in a smart environment. This approach takes into consideration the heterogeneity and the dynamicity of such environments and deals with resource privacy. We propose to describe the available hardware infrastructure and the deployable applications using graphs, and provide a mathematical formalisation of the deployment process based on graph homomorphisms. A decentralised version of a branch and bound graph-matching algorithm is used to find the available hardware entities of the infrastructure that can be used to run the application, respecting its requirements. At last, we describe a goal-directed Multi-Agent System (MAS) for the deployment of applications in ambient systems. We show that the multi-agent paradigm is well-adapted to provide a clear separation between the applicative and the hardware layers, thus increasing resource privacy.

1 Introduction

Ambient Intelligence (AmI) research focuses on the improvement of human interactions with smart applications [12]. These improvements are made possible by the proposal of frameworks and platforms that facilitate the development of context-aware and dynamic applications. However, it is assumed that an underlying interoperable hardware and energy infrastructure already exists [21]. Meanwhile, the Internet of Things (IoT) aims to provide a global infrastructure for the information society, enabling advanced services by interconnecting physical and virtual "things" based on existing and evolving interoperable information and communication technologies [15]. The main challenge of the IoT is to achieve full interoperability of interconnected devices while guaranteeing the trust, privacy and security of communications [3]. However, because of the heterogeneity of such systems, it is difficult to have horizontal communication between connected devices; they cannot communicate directly together. Present applications use devices that are vertically connected, from the device to an external server that collects and processes the data. Moreover, this raises privacy questions: the user does not own his data any more and privacy cannot be guaranteed that way.

© Springer International Publishing AG 2017
C. Badica et al. (eds.), *Intelligent Distributed Computing X*,
Studies in Computational Intelligence 678, DOI 10.1007/978-3-319-48829-5_4

To allow horizontal connections between devices, we need mechanisms that reason on the heterogeneity of systems in order to automatically deploy smart applications provided by AmI on an existing hardware infrastructure provided by the IoT. To address these issues, we propose in Sec. 2 a graph-based model of ambient systems with a mathematical formalisation for the projection of applications on an existing infrastructure. We then present in Sec.3 our multi-agent approach and show why MAS is a well-adapted paradigm to ensure privacy of the resources in smart environments.

2 Modelling the Deployment

2.1 Graph-based Model

Hardware entities of ambient systems are interconnected and used as the support for the running of the applications. These applications are composed of a set of interdependent functionalities that have hardware requirements to be run with a certain quality of service. We use graph theory to model these different entities (hardware, functionalities, requirements), their properties and their relations. Figure 1 shows an example of an application graph. Upper part represents the different functionalities of the application (diamonds), interacting together (dotted arcs). Lower part shows the hardware requirements of the functionalities: the entities (bold rectangles) and the relation between these entities (rounded rectangles). To each node (entity or relation), we can attach properties that characterise it. The available infrastructure is described the same way.

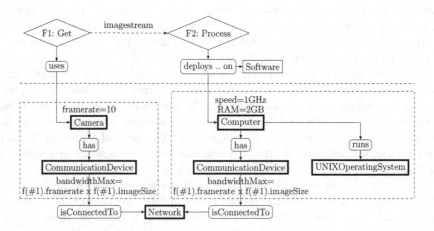

Fig. 1. Example of an application graph

To deploy an application on the hardware infrastructure, we need to find, for each hardware requirement of the application, an available and compatible hardware entity in the infrastructure graph.

2.2 Mathematical Formalisation

We model the deployment by projecting the hardware requirements sub-graph of the application on the infrastructure graph, with graph homomorphisms.

We define P_i as the set of all the possible values of a property i. The property i can be, for example, the bandwidth ($P_{Bandwidth} = \mathbb{R}^+ \cup \{\varnothing, \infty\}$) or the operating system type ($P_{OS} = \{\text{Unix}, \text{Windows}, \text{iOS}, \text{Android}, \varnothing, \infty\}$). For each set P_i, we define a binary operation $\vee_i$ and a partial order $\leqslant_i$. The binary operation $\vee_i$ allows to combine properties, whereas the partial order $\leqslant_i$ determines if the left property value is compatible with the right value. If some nodes of the application sub-graph are projected on the same infrastructure node, then the combination of their properties must respect the partial order with the properties of the infrastructure node. The algebraic structure $(P_i, \vee_i)$ forms a commutative monoid with an absorbing element (∞) and an identity ($\varnothing$), ordered by $\leqslant_i$.

To facilitate the manipulation of properties, we define Π as the Cartesian product of all the property sets: $\Pi = P_1 \times ... \times P_n = \prod_{i=1}^{n} P_i$. An element of this set Π is a tuple of property values. Each node of the graphs is characterised by one of these tuples. We can define a general binary operation $\otimes$ and a general partial order $\circledS$ over Π, defined by:

$$\forall(\pi, \pi') \in \Pi^2 \begin{cases} \pi \otimes \pi' = (p_1 \vee_1 p'_1, \ldots, p_n \vee_n p'_n) \\ \pi \circledS \pi' \Leftrightarrow (p_1 \leqslant_1 p'_1) \wedge \cdots \wedge (p_n \leqslant_n p'_n) \end{cases}$$

We define a graph $G = (V, E, \mathcal{P})$, with V, a set of vertices (entity or relation); E, a set of pairs of vertices corresponding to the edges of the graph; and $\mathcal{P}$, a function from V to Π that associates for each vertex, a vector of properties that characterises the node.

To deploy an application on the infrastructure, we first need to find a projection of the hardware requirements sub-graph of the application on the infrastructure graph. This projection can be seen as a graph homomorphism that associates for each node of the application sub-graph, a compatible node in the infrastructure graph. We define the function $\phi : G \longrightarrow H$. ϕ is an enriched graph homomorphism if and only if:

$$\phi : V_G \longrightarrow V_H \begin{cases} \forall(u, v) \in E_G, (\phi(u), \phi(v)) \in E_H \\ \forall y \in V_H, \bigvee_{x \in \phi^{-1}(y)} \mathcal{P}_G(x) \circledS \mathcal{P}_H(y) \end{cases}$$

For each vertex in the application sub-graph, ϕ attributes a vertex in the hardware infrastructure graph where: the edges between the nodes are respected; and the property combination of all vertices in the application graph that are projected to the same vertex in the hardware infrastructure graph respects the partial order with the properties of this image vertex. Such homomorphism represents a solution for the projection of an application sub-graph on an infrastructure graph. The set of all homomorphisms, $(Hom(G, H))$ represents all the possible solutions for the projection problem. The number of these projections is between 0 and $|V(H)|^{|V(G)|}$. Finding a solution is NP-complete [10].

2.3   Branch and Bound Algorithm

This homomorphism problem can be solved by using an exact graph-matching algorithm. Exact graph-matching refers to different types of problem. The graph homomorphism problem is the weaker form of matching, where the edges must be respected. It is the one we focus on for our problem. The other graph matching problems add new assumptions to the general problem. A graph monomorphism is an injective morphism where each node of the source graph is projected on a different node in the target graph; a graph epimorphism is a surjective morphism where each node of the target graph is the image of one or multiple nodes in the source graph; and a graph (sub-)isomorphism is a bijective morphism between a source graph and a target (sub-)graph.

A lot of algorithms and variants exist to solve these different morphism problems [19]. Probably the most famous is the one from Ullmann [27]. It is a branch and bound algorithm which explores depth-first the source graph and tries to associate for each node of this graph a node in the target graph, respecting the edges. Backtracking is used when an inconsistent state is reached. Some improvements of this tree matching algorithm consist in adding a heuristic for the graph exploration [11]. Another algorithm [18] reformulated the graph isomorphism problem in a constraint problem which can be resolved with a classical CSP engine. Messmer and Bunke present a variant of a matching method for expert systems [20]. This algorithm is based on a recursive decomposition of a graph in sub-graphs. It is particularly efficient for matching a graph with a graph database for which the decompositions could be pre-computed. At last, Babai proposes an algorithm to solve the isomorphism problem in quasipolynomial time [4].

To dynamically deploy an application in a smart environment which is also dynamic, we need an algorithm that can be executed step by step, propose partial solutions and allow the implementation of heuristics to guide the building of the solution. We adapt a classical branch and bound algorithm, like the one proposed by Chein and Mugner [7], by integrating the node properties axiom of our enriched graph-homomorphism. This algorithm takes as inputs the hardware requirement sub-graph G of an application and the infrastructure graph H representing the smart environment. The graph G is depth-first explored and the algorithm tries to successively assign a compatible node in H, following the edges. If there is no solution, then the algorithm backtracks until another compatible node is found in the infrastructure graph. During the execution, the explored nodes in G and their assigned nodes in H represent a partial projection from G to H. When all the nodes in G are assigned, the graph homomorphism problem is solved and a complete projection is found. To get all the possible projections, we just have to memorise a projection and continue the execution of the algorithm by backtracking again.

3   Multi-Agent System (MAS)

Smart environments are characterised by a high dynamicity. The deployment of applications running in such environments has to be adapted to take into consideration their variability (context, devices acquaintances, hardware structure and properties etc.). In real systems, privacy, autonomy, robustness and scalability are essential. That is why we identified MAS as a suitable solution to implement such deployment middleware. Indeed, this paradigm possesses good properties to facilitate local processing of the data, guarantee the autonomy of the different parts of the hardware infrastructure and so handle some aspects of privacy [25]. In this section, we focus on the encapsulation of resource privacy, using agents and agent organisations, to deploy applications.

3.1 Agents and Artifacts

We present the two kinds of component in the approach: agents and artifacts, interacting with each other. The agents are able to reason on the projection of applications with respect to the available infrastructure, whereas artifacts are resources and tools that can be instantiated and/or used by agents in order to interact with the environment and effectively deploy the applications [23]. The former are autonomous and goal-directed while the latter are not.

We propose four classes of agent:

- An *Infrastructure Agent* deals with a part of the global hardware infrastructure. It uses the corresponding graph representation. This one is never shared with other agents. The *Infrastructure Agent* reasons on it, using the algorithm described in the previous section, to propose partial solutions for the deployment of applications. This class of agent has several goals, as it has to: (1) keep the infrastructure graph up to date; (2) propose partial projections of applications, considering the available hardware infrastructure but also the sharing and privacy policy and (3) deploy or undeploy functionalities of applications.
- An *Infrastructure Super Agent* is a representative of a set of *Infrastructure Agents* which are related to it forming a *group*. It acts as a proxy between the agents inside and outside of the group.
- An *Application Agent* manages an entire application during its runtime. It has the graph-based description of the application. The goals of this class of agent are to: (1) guarantee the consistency of the application and (2) deploy or undeploy functionalities of the application if necessary. The *Application Agent* has to interact with several *Infrastructure Agents* in order to deploy the functionalities of the application over the infrastructure.
- At last, the *User Agent* is the interface between the user and the other agents. Through this agent, a user can request the deployment or undeployment of applications.

We also propose two classes of artifact:

- *Deployment artifacts* [14] can be used by the *Infrastructure Agents* in order to effectively deploy some parts of an application, or configure hardware entities so that they can be used by the application.

- *Monitoring artifacts* provide useful contextual information to the deployment software (location of a user, available bandwidth etc.), to help the agents keep their application or infrastructure graph up to date.

The agent decomposition encapsulates a part of the privacy mechanism. Indeed, the graph representation of the available hardware infrastructure managed by an *Infrastructure Agent* is only known by this agent and is never shared with others. Moreover, the architecture helps keep a clear separation between the applicative part, managed by *Application Agents*, and the hardware part, monitored by *Infrastructure Agents*. As agents only have a local view of the system, the privacy is enhanced. At last, *Infrastructure Agents* can be grouped behind an *Infrastructure Super Agent* which, as stated before, acts as a proxy for the agents of this group. From an outside view, this *Infrastructure Super Agent* is seen as a normal *Infrastructure Agent*, resulting in a multi-scale organisation that helps improve privacy. It is then easier to abstract groups of agents and make them invisible from the outside.

3.2   Sharing policies

To improve privacy by controlling the use of resources, we also propose sharing policies. *User Agents* can be authorised, by the owner of some hardware infrastructure, to use some parts of its infrastructure, and cooperate with the associated *Infrastructure Agents* or *Super Agents*, to deploy applications. If a *User Agent* is not authorised by the *Infrastructure (Super) Agent*, it cannot use the hardware resources proposed by this agent. Otherwise, it can have different authorisation levels, from an *Administrator level* that provide a full access to the ressources proposed by the *Infrastructure Agent* to differents *Guest levels* that provide restricted access to resources considered as non critical. These authorisation levels can be modified by the administrators of the *Super Agent*. The *Application Agents* have the same authorisation level as the *User Agent* that created them. They can interact with the authorised *Infrastructure Agents* in order to effectively deploy their application. As the *Application Agents* are created by *User Agents*, they both share the same authorisation level.

In this section, we have shown how privacy is preserved through encapsulation in our MAS. *Infrastructure Agents* keep the information about the hardware infrastructure secret. The *Infrastructure Agent* hierarchy keeps the details of the agent organisation hidden. Privacy policies allow or prevent the sharing of resources to *User Agents*. This results in privacy by design.

3.3   Distributed Deployment Algorithm

The algorithm presented in Sec. 2.3 allows to find the projections of applications on an infrastructure. However, it is centralised and does not take into account important specificities of smart environments, like privacy or scalability. We have seen in the previous paragraphs that our MAS approach tackles this problem; each *Infrastructure Agent* has to deal with a part of the global hardware infrastructure that remains private to ensure privacy. Thus, we have to use a decentralised version of the previous branch and bound algorithm that distribute the representation of the infrastructure graph.

The global infrastructure graph is decomposed in sub-graphs. These sub-graphs are linked through their mutual nodes. Two sub-graphs linked together are adjacent. The way we decide to cut the global infrastructure graph into sub-graphs depends on the application field. For our smart home demonstrator, we decided to cut the sub-graphs at the level of the communication links between two hardware entities, and more specifically at the network nodes. So the sub-graphs are linked together by these different networks that are the ones to be shared. Figure 2 presents an example of a global infrastructure graph. The filled (coloured) nodes represent the nodes where the graph can be cut (e.g. network nodes). The decomposition of this graph into sub-graphs is illustrated in Fig. 3. Each sub-graph can be handled by an *Infrastructure Agent* that never shares this representation and thus keeps the hardware infrastructure secret.

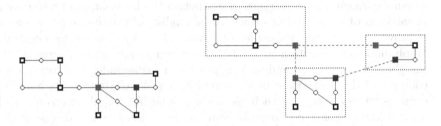

Fig. 2. A global infrastructure graph   **Fig. 3.** Sub-graph decomposition

We can now ask each *Infrastructure agent* to apply the previous algorirthm to locally find a solution to the projection of the application. If there is no local solution, then a partial projection is generated and the agent ask another authorized *Infrastructure agent* in its neibourhood to complete the projection. The partial projections, shared among the agents, indicates only the associated nodes without revealing the details of the infrastructure, thus ensuring privacy.

3.4 Implementation

The multi-agent system has been implemented using a goal-driven approach which helps handle the autonomy and the proactivity of agents [8]. We modelled the agents following the Goal-Plan Separation (GPS) approach [6] which distinguishes between the high level *goal plans* that describe relationships between the goals and low level *action plans* that detail the sequences of concrete actions. Each agent has a main goal plan that describes the top level behaviour, which can be pursued using other goal plans or action plans.

A demonstration version of this MAS has been implemented in a home replica attached to our laboratory. This smart home replica is equiped with commercial connected devices and implements various scenarios applied to home care for dependent persons. The associated applications can be automatically deployed thanks to the demonstration model. This realisation has allowed us to figure out the difficulties of handling the heterogeneity of hardware entities.

4 Related Work

Classical works in AmI propose platforms that offer mechanisms to build context-aware applications [22]. The proposed mechanisms usually handle data and events [16] or wrap hardware/software capabilities into agents [13]. But very few works address the deployment of these applications. Braubach and al. [5] propose a deployment reference model based on a MAS architecture (e.g. agent services) for deploying MAS applications. As an agent is a software entity, the deployment of agents does not have to deal with the high heterogeneity of hardware entities. Some other works in the service-oriented architectures (SOA) community [2] reason on deployment patterns, that specify the structure and constraints of composite solutions on the infrastructure, in order to compose services. Contrary to our approach, the cited paper refers not to the localisation of resources and installation of software, but rather to the binding of existing resources in order to provide the desired composition of services. This is realised using a centralised graph-matching algorithm that takes into account the various requirements for the given service. Flissi and al. [14] propose a meta-model for abstracting the concepts of the deployment of software over a grid. All these works have shortcomings when considering their use for deploying AmI applications on the IoT infrastructure. Some do not take into consideration the heterogeneity of the hardware and software, as well as the interaction between the two layers (i.e. software and hardware). Others do not tackle the privacy problem. And some propose centralised solutions that are not scalable for real life AmI applications.

Privacy in multi-agent systems has already been well explored. Such and al. [25] categorise research on data privacy on different levels: collection, disclosure, processing and dissemination. Multi-agent system specificities have been used to propose different manners of handling the data privacy. Some works focus on norms [17] and privacy policies [26], checked by agent brokers to control the disclosure of the data. Other works [24] use social relationships like trust, intimacy or reputation to select the agents with which data can be shared. Trusted third parties are already used in [9] in order to anonymise the data or the metadata (e.g. IP address, receiver or sender identity), and also to check disclosure authorisations. At last, some works [1] focus on integrating secure communication in the agent platforms by using well known encryption protocols. All these works use MAS in order to provide data privacy. In our work, as explained in Sec. 3, we merely take advantage of MAS properties to handle the privacy of the hardware resources and of the structure of the system.

5 Conclusion

In this paper, we presented a multi-agent approach for the deployment of intelligent applications in smart environments. We proposed to handle the heterogeneity of ambient systems by describing the available hardware infrastructure and the deployable applications using graphs. To deploy an application, we needed first to find the projection of the hardware requirements of an application graph on the infrastructure graph. This projection was modelled as a graph homomorphism. A centralized branch and bound graph-matching algorithm was used as a base. This algorithm was then improved to allow the distribution of the infrastructure graph and thus promote a local-search first. At last, this algorithm was encapsulated in a Multi Agent System that improved resource privacy. The implementation was carried out using a goal-oriented approach and a demonstrator of the deployment solution was applied in a real Smart Home replica. The next steps of this ongoing work will be to evaluate the performances of this approach by varying the structure of the graphs and the agent organisation. We also would like to consider application data privacy by defining data privacy policies in order to facilitate the local processing and storage of the data. The user should decide which kind of data he authorises to come out of his home infrastructure. This will impact the reasoning on the deployment; the hardware entities have to be chosen in order to fulfil this new data privacy policy.

References

1. Alberola, J., Such, J., Garcia-Fornes, A., Espinosa, A., Botti, V.: A performance evaluation of three multiagent platforms. Artificial Intelligence Review 34(2), 145–176 (2010)
2. Arnold, W., Eilam, T., Kalantar, M., Konstantinou, A., Totok, A.: Automatic realization of soa deployment patterns in distributed environments. In: Bouguettaya, A., Krueger, I., Margaria, T. (eds.) Service-Oriented Computing ICSOC 2008, Lecture Notes in Computer Science, vol. 5364, pp. 162–179. Springer Berlin Heidelberg (2008)
3. Atzori, L., Iera, A., Morabito, G.: The internet of things: A survey. Computer Networks 54(15), 2787 – 2805 (2010)
4. Babai, L.: Graph isomorphism in quasipolynomial time. CoRR abs/1512.03547 (2015)
5. Braubach, L., Pokahr, A., Bade, D., Krempels, K.H., Lamersdorf, W.: Deployment of distributed multi-agent systems. In: Gleizes, M.P., Omicini, A., Zambonelli, F. (eds.) Engineering Societies in the Agents World V, Lecture Notes in Computer Science, vol. 3451, pp. 261–276. Springer Berlin Heidelberg (2005)
6. Caval, C., El Fallah Seghrouchni, A., Taillibert, P.: Keeping a clear separation between goals and plans. In: Dalpiaz, F., Dix, J., van Riemsdijk, M. (eds.) Engineering Multi-Agent Systems, Lecture Notes in Computer Science, vol. 8758, pp. 15–39. Springer International Publishing (2014)
7. Chein, M., Mugnier, M.L.: Graph-based Knowledge Representation: Computational Foundations of Conceptual Graphs. Springer, London (2008)
8. Cheong, C., Winikoff, M.: Agent-Oriented Software Engineering VI: 6th International Workshop, AOSE 2005, Utrecht, The Netherlands, July 25, 2005. Revised and Invited Papers, chap. Hermes: Designing Goal-Oriented Agent Interactions, pp. 16–27. Springer Berlin Heidelberg, Berlin, Heidelberg (2006)

9. Cissée, R., Albayrak, S.: An agent-based approach for privacy-preserving recommender systems. In: Proceedings of the 6th International Joint Conference on Autonomous Agents and Multiagent Systems. pp. 182:1–182:8. AAMAS '07, ACM, New York, NY, USA (2007)

10. Conte, D., Foggia, P., Sansone, C., Vento, M.: Thirty years of graph matching in pattern recognition. IJPRAI 18(3), 265–298 (2004)

11. Cordella, L.P., Foggia, P., Sansone, C., Vento, M.: An improved algorithm for matching large graphs. In: In: 3rd IAPR-TC15 Workshop on Graph-based Representations in Pattern Recognition, Cuen. pp. 149–159 (2001)

12. Ducatel, K., Bogdanowicz, M., Scapolo, F., Leijten, J., Burgelman, J.: Scenarios for ambient intelligence in 2010 (2001)

13. El Fallah Seghrouchni, A., Olaru, A., Nguyen, N.T.T., Salomone, D.: Ao dai: Agent oriented design for ambient intelligence. In: Desai, N., Liu, A., Winikoff, M. (eds.) PRIMA. Lecture Notes in Computer Science, vol. 7057, pp. 259–269. Springer (2010)

14. Flissi, A., Dubus, J., Dolet, N., Merle, P.: Deploying on the grid with deployware. In: Cluster Computing and the Grid, 2008. CCGRID '08. 8th IEEE International Symposium on. pp. 177–184 (2008)

15. ITU-T: Overview of the internet of things, recommendations (2012)

16. Johanson, B., Fox, A., Winograd, T.: The interactive workspaces project: Experiences with ubiquitous computing rooms. IEEE Pervasive Computing 1(2) (2002)

17. Krupa, Y., Vercouter, L.: Contextual integrity and privacy enforcing norms for virtual communities. In: Boissier, O., El Fallah Seghrouchni, A., Hassas, S., Maudet, N. (eds.) MALLOW. CEUR Workshop Proceedings, vol. 627. CEUR-WS.org (2010)

18. Larrosa, J., Valiente, G.: Constraint satisfaction algorithms for graph pattern matching. Mathematical Structures in Computer Science 12, 403–422 (2004)

19. McKay, B.D., Piperno, A.: Practical graph isomorphism, ii. Journal of Symbolic Computation 60, 94–112 (2014)

20. Messmer, B.T., Bunke, H.: Efficient subgraph isomorphism detection: A decomposition approach. IEEE Trans. on Knowl. and Data Eng. 12(2) (2000)

21. O'Hare, G.M.P., Collier, R., Dragone, M., O'Grady, M.J., Muldoon, C., de J. Montoya, A.: Embedding agents within ambient intelligent applications. In: Bosse, T. (ed.) Agents and Ambient Intelligence, Ambient Intelligence and Smart Environments, vol. 12, pp. 119–133. IOS Press (2012)

22. Olaru, A., Florea, A.M., Seghrouchni, A.E.F.: A context-aware multi-agent system as a middleware for ambient intelligence. Mobile Networks and Applications 18(3), 429–443 (2013)

23. Ricci, A.: Agents and coordination artifacts for feature engineering. In: Ryan, M.D., Meyer, J.J.C., Ehrich, H.D. (eds.) Objects, Agents, and Features. Lecture Notes in Computer Science, vol. 2975, pp. 209–226. Springer (2003)

24. Such, J.M., Espinosa, A., GarcíA-Fornes, A., Sierra, C.: Self-disclosure decision making based on intimacy and privacy. Inf. Sci. 211, 93–111 (Nov 2012)

25. Such, J.M., Espinosa, A., Garca-Fornes, A.: A survey of privacy in multi-agent systems. The Knowledge Engineering Review 29, 314–344 (2014)

26. Udupi, Y.B., Singh, M.P.: Agents and peer-to-peer computing. chap. Information Sharing Among Autonomous Agents in Referral Networks, pp. 13–26. Springer-Verlag (2010)

27. Ullmann, J.R.: An algorithm for subgraph isomorphism. J. ACM 23(1), 31–42 (1976)

A Guidance of Ambient Agents Adapted to Opportunistic Situations

Ahmed-Chawki Chaouche[1], Jean-Michel Ilié[2] and Djamel Eddine Saïdouni[1]

[1] MISC Laboratory, University Abdelhamid Mehri - Constantine 2,
Ali Mendjeli Campus, 25000 Constantine, Algeria
{ahmed.chaouche,djamel.saidouni}@univ-constantine2.dz
[2] LIP6, UMR 7606 UPMC - CNRS,
4 Place Jussieu, 75005 Paris, France
jean-michel.ilie@upmc.fr

Abstract. In this paper, we address ambient systems whose computational process is based on autonomous and context-aware intelligent agents. The planning management framework we propose is agent-centered and looks for an efficient guidance improving the satisfaction of the agent's intentions, taken as a whole. This also includes the intentions pushed by some opportunistic situations, according to their relevances for the agents. Originally, the formal model we propose allows to dynamically schedule the plans of the intentions concurrently, allowing to extract the traces having a maximum relevance coding.

Keywords: Ambient system, context-awareness, BDI agent, planning guidance, opportunistic situations.

1 Introduction

Multi-agent systems (MAS) offer interesting frameworks for the development of ambient intelligence (AmI). Actually, their agents are considered as intelligent, proactive and autonomous, able to adapt to the changes of contexts [1, 2, 3]. Many modern applications are concerned, in particular, applications where AmI agents are strongly connected with humans with the aim to assist them, contributing to define smart environments.

At the agent level, general consensus exists that the Belief-Desire-Intention (BDI) model is well suited for describing an agents mental state. Rich expressivity is obtained by considering types of goals to characterize the possible desires, e.g. in [4, 5]. Anyway, intentions are no more that some of the specified goals, whose action plans can be executed by the agent. When searching the next action to perform, plans are often managed separately one from the other, e.g. in [6, 7], but the recent work of [8] demonstrates how a planner can execute the (intention) plans concurrently, based on scheduling information over the specified intentions.

In ambient environment, it is challenging to conceive highly reactive guidance mechanism, which allows the agent to adapt its behavior on-the-fly, In many works, the guidance process is reduced to better selecting a sequence of actions for the agent. The context of the agent is usually taken into account dynamically as in [6, 7] and some utility concepts can be injected in the selection process to better correspond to the userss desires [9]. Recently, the work of [8, 10] demonstrates even more efficiency in proposing an action scheduling technique over the so-called intention plans, that are the plans designed to achieve the current intentions of the agent. This is used to formally maximize the achievement of the set of the agent intentions, in regards to the expected context changes.

In this paper, we aim at providing an efficient guidance mechanism taking opportunistic situations into account. Accordingly to [11], we know that an agent can consider opportunistic situations provided that some goals are specified for that, namely long-term goals. As in [5], this type of goals can be suspended or in contrast activated to take opportunistic situations into account, however, up to our knowledge, nothing is said about the impact on the existing intentions since this pertains to the (sometimes complex) deliberation process of the agent. To allow maximizing the intentions, we propose to abstract this process so as to be sufficiently informed on the relative importance of the intentions, in between them. This forces us to reconsider the scheduling technique of [8], to dynamically handle opportunistic intention plans whenever this is possible.

The remaining of this paper is organized as follows: Section 2 highlights a realistic travel scenario given as an illustration of the concepts proposed in this paper. In Section 3, we introduce the concepts of our opportunistic planner, based on a formal language, namely *AgLOTOS*, expressing the possible execution of intention plans in a compact and concurrent way. In Section 4, a contextual planning management is automatically derived, keeping inspiration from the Contextual Planning System (CPS) semantical work of [8] but augmented to considerations related to the relevances of intention plans. In Section 5, we demonstrate how the proposed approach accepts replanning, focusing on the fact that opportunistic situations can occur. The last section concludes this paper and brings out our very next perspectives.

2   An Ambient Scenario with Opportunistic Situations

The following travel scenario takes place on an airport representing an AmI system. Any entering traveler follows different successive steps, like *check-in* and *boarding*. This generates different intentions for the agent, and for each, different possible alternative plans in order to be achieved. In particular, there may be two ways to realize the check-in, either by using a self-serve check-in system or helped by a human assistant whenever the registering of specific bags is required.

Opportunistically, this traveler aims at buying a newspaper, speaking about his destination country. In fact, he wants to be aware of events which currently occur in this country, as soon as possible. Such newspaper surely exists within the airport, either before or after getting through the customs control. In any case, the main objective of the traveler is not to miss the plane.

For a better guidance, the airport assistance provides to each traveler a smart application to closely assist him. The smart device of each traveler is then equipped with some software agent, so that the Graphical User Interface allows the traveler to select intentions and preferences, from the possible desires that can be enabled in the airport. Here, the check-in and the boarding are the important intentions for the traveler and must be achieved in sequence, whereas the intention "buying the newspaper" is assumed to be achieved opportunistically, that means without much planning effort in-between the check-in and the boarding stages and without challenging the achievements of the important intentions.

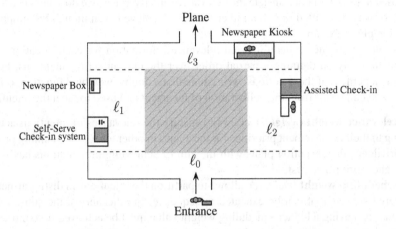

Fig. 1. Spatial view of the considered airport

The modeling of this scenario is enhanced in Sections 3 and 4, distinguishing among the relative importances of the travelers intentions. As described in Figure 1, the airport is split in four distinct areas wherein the traveler behaves: ℓ_0 is the airport entrance, ℓ_1 contains the self-serve check-in system and a newspaper-box, ℓ_2 corresponds to the assisted check-in and ℓ_3 to the boarding area, within which there is a newspaper kiosk (newsstand).

3   A Concurrent and Opportunistic Planner

Traditionally the mental process of a BDI agent is able to deliberate and compute a set of intentions at real time. These intentions are selected goals that can be executed in a concurrent way. As this property is not easy to check in AmI system, we adopt the approach developed in [8] which introduces a specific planning process able to analyze the effect of scheduling the current set of intentions (at the action level). This planning process can return a *(global) execution* plan for the agent, that maximizes the concurrent achievement of the set of intentions (see section 4). This is realized in regard to the evolution of the current context of the agent, as consequences of performing actions in plans. It should be noticed that the perfomance of an action generally requires contextual preconditions and should impact the agent context, hence can prohibit the performance of further actions. By default, the set of intention plans of the agent are assumed to be performed concurrently whenever this is possible, however the mental process can specify a static partial order of commitments among the intentions. Actually, this overpasses the traditional view of intentions, but can be an easy way to solve some conflict or merely to functionnaly order some of them.

In practice, an intention which is triggered by an opportunistic situation is no more than a classical intention, however, there is an underlying notion of relevance of this intention with respect to the other ones. Unfortunately, the scheduling of intention plans proposed in [8] could not translate this relevance directly. relevance does not necessarily stand for being executed first, but rather for being achieved even though less important intention plans are run.

We now introduce scheduling and relevance information for each intention. Both information rely on defining a partial order over the current set of intentions. For an easy management of these notions, we assume that a double weighted function *weight* : $I \rightarrow \mathbb{N} \times \mathbb{N}$ can result from the deliberation of the agent and used to label the intentions:

The relevance weight (*weight$_r$*) allows to specify the relevances of intentions according to their achievement priorities for the agent. In other terms, the planning process privileges the execution plans with the aim of achieving the intentions having the highest priorities, first.

The scheduling weight (*weight$_s$*) allows to partition the intentions in distinguished ordered subsets that can be executed in sequence. In other terms, the subset of intentions having a higher scheduling weight value must be achieved first, moreover, two intentions of the same subset can be achieved concurrently, independently of any relevance weights.

The current set of weighted intentions is denoted I^w, $weight_r(i)$ represents the relevance weight of the intention i and $weight_s(i)$ represents the scheduling weight of i. In regard to any intention i, the corresponding weighted intention is denoted $i^{(r,s)}$, where $r = weight_r(i)$ and $s = weight_s(i)$. Consider in addition that 0 represents the lowest value of any weight system.

In the concern of our scenario, we have $I^w = \left\{ i_c^{(1,1)}, i_b^{(1,0)}, i_y^{(0,0)} \right\}$. Observe that the subscripts indicate the meanings of each intention: 'b' stands for boarding, 'c' for checking, and 'y' for buying. It is clear that $i_c^{(1,1)}$ and $i_b^{(1,0)}$ are more relevant than $i_y^{(0,0)}$, moreover, $i_c^{(1,1)}$ must be achieved before the two others.

We next show how the scheduling weights are used to build the so-called *agent plan*, which is an algebraic expression characterizing all the possible execution plans, with respect to the current set of intentions of the agent and the scheduling weights (see Section 3.1). In contrast, the relevance weights are taken into account when the execution plans are finally derived from the agent plan, w.r.t. the current context of the agent (see Section 4).

3.1 Agent Plan Structure

An agent plan statically specifies the way to compose the intention plans associated with the set of intentions. For clarity reasons, let us explain its structure from an abstract graphical representation called the *agent plan structure*. This structure looks like a tree highlighting that the *agent plan* ($\overline{P}$) is composed of *intention plans* ($\widehat{P}$), that

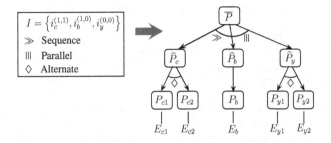

Fig. 2. The agent plan structure of the travel scenario

means the plans to realize the intentions, knowing in addition that each of the intention plans is often an alternate of *elementary plans*. For instance, the agent plan structure of our scenario is described in Figure 2. It brings out three intention plans $\widehat{P_b}$, $\widehat{P_c}$ and $\widehat{P_y}$, concretizing the three intentions of agent. At a lower level, $\widehat{P_b}$ is specified from one elementary plan P_b, whereas the other two intention plans $\widehat{P_c}$ and $\widehat{P_y}$ are composed of two alternative elementary plans, respectively, P_{c1}, P_{c2} and P_{y1}, P_{y2}.

An agent plan can be compactly represented by introducing algebraic operators to compose the different kind of plans. As mentioned in the legend of Figure 2, three composition operators are introduced: '|||' is the parallel composition of intentions plans whereas '≫' is the sequential composition. Moreover, '◊' is the alternation operator of elementary plans. Formally,

$$\overline{P} ::= \widehat{P} \mid \overline{P}|||\overline{P} \mid \overline{P} \gg \overline{P}$$
$$\widehat{P} ::= P \mid \widehat{P} \Diamond \widehat{P}$$

The algebraic expression of an agent plan can be automatically deduced from the set I of intentions and the associated *weights* function, according to the following three functions:

- *libp* $: \mathcal{I} \rightarrow 2^{\mathcal{P}}$, features the library of elementary plans. This yields, for each intention $i \in \mathcal{I}$, a set of instantiated elementary plans dedicated for achieving i.
- *options* $: \mathcal{I} \rightarrow \widehat{\mathcal{P}}$, yields for any $i \in \mathcal{I}$, an intention plan of the form: $\widehat{P_i} = \Diamond_{P \in libp(i)} P$.
- *plan* $: 2^I \rightarrow \overline{\mathcal{P}}$, creates the final agent plan $\overline{P}$ according to I. In fact, depending on how the intentions are weighted by *weights*, the intention plans $(\widehat{P_i} = options(i))$ are composed using either parallel operator or sequential one. This yields $\widehat{P_i}|||\widehat{P_j}$ whenever $weights(i) = weights(j)$, and $\widehat{P_i} \gg \widehat{P_j}$, whenever $weights(i) > weights(j)$.

Applying these functions results in sequences of bracketed sub-expressions representing each one the part of the intentions plans having a same scheduling weight. Let consider the weighted intention set of the scenario, $I^w = \left\{ i_c^{(1,1)}, i_b^{(1,0)}, i_y^{(0,0)} \right\}$, and the corresponding plan structure illustrated in Figure 2. The agent plan $\overline{P}$ is built by composing the intention plans $\widehat{P_c}, \widehat{P_b}$ and $\widehat{P_y}$, as follows: $\overline{P} ::= \widehat{P_c} \gg (\widehat{P_b}|||\widehat{P_y})$, where $\widehat{P_c} ::= P_{c1} \Diamond P_{c2}$, $\widehat{P_b} ::= P_b$ and $\widehat{P_y} ::= P_{y1} \Diamond P_{y2}$. Further, each elementary plan P_k is associated with an expression E_k describing its behavior, as presented in Table 1.

Table 1. Part of LibP library corresponding to the scenario

$I_{CheckIn}$	E_{c1} ::= $move(\ell_1); selfReg(\ell_1); exit$
	E_{c2} ::= $move(\ell_2); assistedReg(\ell_2); exit$
$I_{Boarding}$	E_b ::= $move(\ell_3); board(\ell_3); exit$
I_{Buying}	E_{y1} ::= $buy(\ell_1); exit$
	E_{y2} ::= $buy(\ell_3); exit$

Further, each elementary plan P_k is associated with an expression E_k representing its behavior. Focusing on our scenario, Table 1 highlights an instantiation of the library of plans of the agent. It is indexed by the possible intentions. In [8], the elementary expressions are formed in a LOTOS-like way, allowing rich constructions formed of concurrent processes. In this paper, the expressions are kept simple, by only using the *prefixing operator* ; to perform actions in sequence. Actions are instantiated to deal with the context of the airport. They are situated, for instance, $move(\ell_1)$ stands for moving to location ℓ_1 and $buy(\ell_3)$ stands for buying the newspaper at location ℓ_3.

3.2 Algebraic Configuration of an Agent Plan

The state of an agent plan $\overline{P}$, also called an *agent plan configuration*, is denoted $[\overline{P}]$. Its expression is fundamentally based on the agent plan structure presented in this section, and on the fact that each elementary plan is known to be associated with an expression representing its behavior.

The *canonical rules* of Definition 1, specify how $[\overline{P}]$ is formed compositionally from some *intention plan configurations*, like $(E, \widehat{P})$ (*rule 1*), themselves built from an alternate of *elementary plan configurations*, like (E_k, P_k) (*rule 2*) which represents an elementary plan identified by P_k, and its behavior expression is E_k.

For sake of simplicity in this paper, the semantics of the alternate operator is reduced to the simple non-deterministic choice, standard in LOTOS: $\lozenge^{k=1..n}E_k \equiv [\,]^{k=1..n}E_k$. We will see that this allows to test every elementary plan for achieving the corresponding intention plan, in different execution plan (see Section 4). Continuing the former example, we have $[\overline{P}] = (E_{c1}[\,]E_{c2}, \widehat{P}_c) \gg \Big((E_b, \widehat{P}_b) ||| (E_{y1}[\,]E_{y2}, \widehat{P}_y) \Big)$.

Definition 1. *(Generic representation of an agent plan configuration)*
Any Agent plan configuration $[\overline{P}]$ has a canonical representation defined by the following two rules:

$$(1) \quad \frac{\overline{P}::=\widehat{P} \qquad \widehat{P}::=\lozenge^{k=1..n}P_k \qquad P_k::=E_k}{[\overline{P}]::=([\,]^{k=1..n}E_k, \widehat{P})} \qquad (2) \quad \frac{\overline{P}::=\overline{P_1} \odot \overline{P_2} \qquad \odot \in \{|||, \gg\}}{[\overline{P}]::=[\overline{P_1}] \odot [\overline{P_2}]}$$

4 Contextual Planning Guidance

In this section, we show how to build a *Contextual Planning System*, denoted *CPS*. It is a transition system representing all the possible evolutions of the agent plan in terms of actions. With respect to some evolution, a *contextual planning state* takes into account some contextual information, here the agent location, and also an information on the intention plans to be achieved.

Definition 2. *(Contextual planning state)*
A contextual planning state is a tuple (ps, ℓ, T)*, where ps is any planning state,* ℓ *corresponds to a contextual information, here restricted to a location for the agent, and T is the subset of intention plans which are terminated.*

4.1 Contextual Planning System

With respect to the intention set I, a CPS is built by using the semantic rules of AgLOTOS. From any contextual planning state, the semantics yields the actions that can be offered and also, for each one, the resulting contextual planning states.

Definition 3. *(Contextual Planning System)*
The Contextual Planning System (CPS for short) is a labeled Kripke structure $\langle S, s_0, Tr, \mathcal{L}, \mathcal{T} \rangle$ *where:*

- *S is the set of contextual planning states,*
- $s_0 = (ps, \ell, \emptyset) \in S$ *is the initial contextual planning state,*
- $Tr \subseteq S \times Act \times S$ *is the set of transitions. The transitions are denoted* $s \xrightarrow{a} s'$ *such that* $s, s' \in S$ *and* $a \in Act$*,*
- $\mathcal{L} : S \to \Theta$ *is the location labeling function*
- $\mathcal{T} : S \to 2^{\widehat{P}}$ *is the termination labeling function which captures the terminated intention plans.*

The CPS is built from an initial contextual planning state, e.g. $([\overline{P}], \ell, \emptyset)$, such that $[\overline{P}]$ is the initial agent plan configuration and ℓ is the location ℓ currently considered for the agent. At that point, all the intention plans mentioned in $[\overline{P}]$ are specified not being achieved (i.e. $T = \emptyset$).

In a CPS, any transition $s \xrightarrow{a} s'$ represents an action to be performed. Like in the STRIPS description language [7], the actions are associated with preconditions and effects. In our approach, the preconditions only concern the contextual information attached to the source state. Let $pre(a)$ be the precondition of any action a, e.g. $pre(a(\ell)) = \ell = \mathcal{L}(s)$.

The CPS of traveler agent, is illustrated in Figure 3. It is built from the initial CPS state, $s_0 = ([\overline{P}], \ell_0, \emptyset)$, taking into account the current location ℓ_0 of the traveler. Observe that the actions which are not realizable are represented by dashed transitions, e.g. from the states s_7, s_{10} and s_{15}. From these states, $pre(buy(\ell_1)) = \ell_1 \neq \mathcal{L}(s)$.

Each trace from s_0 in the CPS supports a possible execution plan corresponding to the sequence of mentioned actions. An example of execution plan, achieving all the agent intentions is given below. It expresses that the traveler can order the check-in using a self-serve check-in system, the buying of a newspaper from the newspaper-Box in ℓ_1, and finally the boarding in plane from ℓ_3:

$$\left((E_{c1}[\,]E_{c1}, \widehat{P_c}) \gg ((E_b, \widehat{P_b}) ||| (E_{y1}[\,]E_{y2}, \widehat{P_y})), \ell_0, \emptyset \right) \xrightarrow{move(\ell_1)}$$

$$\left((E'_{c1}, \widehat{P_c}) \gg ((E_b, \widehat{P_b}) ||| (E_{y1}[\,]E_{y2}, \widehat{P_y})), \ell_1, \emptyset \right) \xrightarrow{selfReg}$$

$$\left((E''_{c1}, \widehat{P_c}) \gg ((E_b, \widehat{P_b}) \| (E_{y1}[\,]E_{y2}, \widehat{P_y})), \ell_1, \emptyset\right) \xrightarrow{\tau_{c1}}$$
$$\left((E_b, \widehat{P_b}) \| (E_{y1}[\,]E_{y2}, \widehat{P_y}), \ell_1, \{\widehat{P_c}\}\right) \xrightarrow{buy(\ell_1)} \left((E_b, \widehat{P_b}) \| (E'_{y1}, \widehat{P_y}), \ell_1, \{\widehat{P_c}\}\right) \xrightarrow{\tau_{y1}}$$
$$\left((E_b, \widehat{P_b}), \ell_1, \{\widehat{P_c}, \widehat{P_y}\}\right) \xrightarrow{move(\ell_3)} \left((E'_b, \widehat{P_b}), \ell_3, \{\widehat{P_c}, \widehat{P_y}\}\right) \xrightarrow{board}$$
$$\left((E''_b, \widehat{P_b}), \ell_3, \{\widehat{P_c}, \widehat{P_y}\}\right) \xrightarrow{\tau_b} \left(stop, \ell_3, \{\widehat{P_c}, \widehat{P_b}, \widehat{P_y}\}\right)$$

4.2 Guidance Analysis from the CPS

In order to guide the agent efficiently, the CPS guidance, previously proposed in [8], selects an execution trace which maximizes the number of intentions that can be achieved. This can be captured over the set $\Sigma \subseteq 2^{Tr}$ of all the possible traces (σ) of the CPS by an analysis of the set $end(\sigma)$ corresponding to the terminated intention plans obtained at the end of σ.

In this paper, we adapt this notion in a more abstract way focusing on the relevance weights of the intentions. With respect to any trace σ, we are able to extend the $weight_r$ notion to the trace. Actually, this weight can be evaluated for each terminated intention, specified in $end(\sigma)$. This allows us to build a unique word of relevance for each trace, denoted $weight_r(\sigma)$, composed of the relevance weights of the terminated intentions of $end(\sigma)$. In order to compare the traces of the CPS, every word is normalized knowing the cardinality $|I|$ of the set of intentions, by possibly adding a series of 0 weight on the right of words up to obtain a word of length $|I|$, moreover, the weights are placed in the word according to a decreasing order.

Definition 4. *Consider any two intentions i and j such that $weight_r(i) < weight_r(j)$, a trace σ of the CPS is relevant iff the fact that $\widehat{P_i}$ belongs $end(\sigma)$ implies that $\widehat{P_j}$ is also in $end(\sigma)$. An optimal relevant trace is defined over the relevant traces of the CPS as a trace σ having the maximum normalized value $weight_r(\sigma)$.*

For instance considering a set of three intentions and its associated CPS, a trace σ_1 is relevant w.r.t. another trace σ_2 in case their respective relevance weights are $weight_r(\sigma_1) = 33$ (s.t. 33 is normalized in 330) and $weight_r(\sigma_2) = 321$.

Corresponding to our scenario, the CPS of Figure 3 contains 7 optimal relevant traces over 9 relevant traces. For instance, the trace carried out by $s_0 \rightarrow s_1 \rightarrow s_3 \rightarrow s_5 \rightarrow s_8 \rightarrow s_{12} \rightarrow s_{17} \rightarrow s_{22} \rightarrow s_{25}$ is optimal relevant, whereas $s_0 \rightarrow s_1 \rightarrow s_3 \rightarrow s_5 \rightarrow s_7 \rightarrow s_{10} \rightarrow s_{15}$ is just relevant.

5 Replanning: The Opportunistic Case

In the standard case, the AmI agent can call its mental process to build a CPS and yield an optimal relevant trace. The execution plan supported by such a trace is performed state by state, since this corresponds to maximizing the achievements of the set of intentions taken as reference.

Taking an opportunistic situation into account can lead to an update of the weighted set of intentions. This causes the need to search a new execution plan from the current

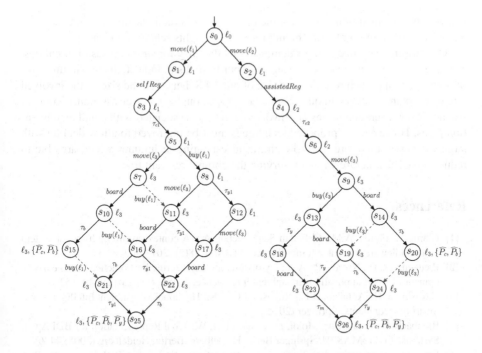

Fig. 3. The CPS corresponding to the traveler agent plan

context of the agent. Then, an agent plan is built, keeping in mind that the intention plans that are finally maintained in the set of intentions may be already partially executed. Hopefully, the modular approach of AgLOTOS allows exploiting the planning structure of the agent plan, providing so a direct way to make the revision of an agent plan. The expressions to be associated with the maintained intentions plans are in fact those extracted from the last planning state reached by the agent. Lastly, a CPS is built from which an optimal relevant trace is brought out.

6 Conclusion

This paper reuses the theoretical advances of the AgLOTOS framework which appears to be a powerful way to express any agent plan as a compact expression of concurrent processes. In this framework, taking into account an opportunistic situation relies first on a partial replanning of the intention plans, together with the specification of a relevance weight for each intention.

This forces us to propose a new management for the CPS which is the contextual action model of the AgLOTOS framework. Among the traces brought out by the CPS, optimal relevant traces are now searched from an analysis of the intention termination sets of the traces. Thus, the agent can be guided to maximize the achievements of its intentions, but the intention plans of some opportunistic situations are taken into account whenever the intention plans of higher relevance weights remain achievable.

Our current experimentations demonstrate efficiency for simple cases of intentions, as in the traveler scenario used to exemplify our formalism. Indeed, the maximum complexity of our approach relies on the size of one CPS, hence on the size of the involved intentions plans. Our technique seems then operational for a user assistance context, whenever few users intentions are involved at a time, associated with small-size intention plans. However, the proposed techniques must be improved to allow dealing with large use cases, therefore, we now aim at investigating techniques which are able to reduce the CPS structure while preserving the optimal relevant traces.

References

[1] Olaru, A., Florea, A.M., El Fallah Seghrouchni, A.: A context-aware multi-agent system as a middleware for ambient intelligence. MONET **18**(3) (2013) 429–443
[2] Preuveneers, D., Novais, P.: A survey of software engineering best practices for the development of smart applications in ambient intelligence. JAISE **4**(3) (2012) 149–162
[3] Nakashima, H., Aghajan, H.K., Augusto, J.C., eds.: Handbook of Ambient Intelligence and Smart Environments. Springer (2010)
[4] Braubach, L., Pokahr, A., Moldt, D., Lamersdorf, W.: Goal Representation for BDI Agent Systems. In: ProMAS'04. Springer Berlin Heidelberg, Berlin, Heidelberg (2005) 44–65
[5] Braubach, L., Pokahr, A.: Representing Long-Term and Interest BDI Goals. In: ProMAS'09. Springer Berlin Heidelberg, Berlin, Heidelberg (2010) 201–218
[6] Singh, D., Sardina, S., Padgham, L., Airiau, S.: Learning context conditions for BDI plan selection. In: AAMAS, IFAAMAS (May 2010) 325–332
[7] Meneguzzi, F., Zorzo, A.F., da Costa Móra, M., Luck, M.: Incorporating planning into BDI agents. Scalable Computing: Practice and Experience **8** (2007) 15–28
[8] Chaouche, A.C., El Fallah Seghrouchni, A., Ilié, J.M., Saïdouni, D.E.: A Higher-order Agent Model with Contextual Management for Ambient Systems. In: TCCI XVI. Volume 8780 of LNCS. Springer Berlin Heidelberg (2014) 146–169
[9] Nunes, I., Luck, M.: Softgoal-based plan selection in model-driven bdi agents. AAMAS'14 (2014) 749–756
[10] Chaouche, A.C., El Fallah Seghrouchni, A., Ilié, J.M., Saïdouni, D.: Improving the contextual selection of bdi plans by incorporating situated experiments. In Chbeir, R., Manolopoulos, Y., Maglogiannis, I., Alhajj, R., eds.: Artificial Intelligence Applications and Innovations. Volume 458 of IFIP Advances in Information and Communication Technology. Springer International Publishing (September 2015) 266–281
[11] Isenberg, D.: The tactics of strategic opportunism. Harvard Business Review **87** (1987) 92–97
[12] Chaouche, A.C., El Fallah Seghrouchni, A., Ilié, J.M., Saïdouni, D.E.: A Dynamical Plan Revising for Ambient Systems. Procedia Computer Science **32** (June 2014) 37 – 44
[13] Boukharrou, R., Chaouche, A.C., El Fallah Seghrouchni, A., Ilié, J.M., Saïdouni, D.E.: Dealing with temporal failure in ambient systems: A dynamic revision of plans. Journal of Ambient Intelligence and Humanized Computing **6**(3) (2015) 325–336

Extended Context Patterns – A Visual Language for Context-Aware Applications*

Andrei Olaru[1] and Adina Magda Florea[1]

Computer Science and Engineering Department
University Politehnica of Bucharest
060042 Bucharest, Romania
cs@andreiolaru.ro, adina.florea@cs.pub.ro

Abstract. The paper presents a visual language that can help users of a context-aware application represent the current situation, or situations they wish detected, in a language that is both formally defined, and readable and understandable by humans and machines alike. Inspired from Regular Expressions, the concept of Extended Concept Pattern provides both conciseness and expressive power, allowing for specifying negation, and for indicating repeating or alternative structures.

Keywords: Artificial intelligence, Ambient intelligence, Context-awareness, Software agents, Graph theory

1 Introduction

Context-aware applications [11] are gaining a great deal of traction today, as the use of context data enables an application to appear to the user as smart and useful, its actions making sense in the current situation [5,1]. Currently, however, context-awareness is (a) *programmed* and, more than that, (b) *pre-programmed* in the applications. Let us explain.

First, context-aware actions or rules are usually embedded in the code of the application, and when they are not, they are represented in a language that is machine-oriented and understandable only by programmers, rather than readable by the user of the application. Second, the user is unable to change the behavior of the context-aware application, such that in some situations the application reacts differently than pre-programmed, but closer to the desires of the user. In part, this is because the user is generally unable to easily understand and modify the context-aware behavior of the application.

This paper introduces a visual language for the representation of context-based behavior. It allows the user to work with the representation of context and to express situations that should be detected and actions that should be taken. The constructs offered by the language offer features of increasing complexity, which correspond to an increasing formal preparation that is required

* This work has been supported by the grant *CAMI: Artificially intelligent ecosystem for self-management and sustainable quality of life in AAL*, AAL Programme, 2015-2018.

to understand how they work. However, usage at lower level should be available to the great majority of users.

The language that we present in this paper is that of *Extended Context Patterns*. It uses as background the formal representations of *Context Graphs* and *Context Patterns*, which have been introduced in previous work to represent the current situation of the user and situations that are desired to be detected [10]. An Extended Context Pattern allows a developer or a user to better understand, use and modify the representation of a situation.

The formalism of Extended Context Patterns is related to the textual linear graph representation that we have previously developed, and improves Context Patterns by giving them more expressive power and making them easier to use. Taking inspiration from Regular Expressions, Extended Context Patterns include *operators* such as transitive closure, alternation, and negation, increasing their power of representation.

Throughout the paper, we will use the following running example: Joe has an elderly mother, Emily, who lives alone. Emily is part of an Ambient Assisted Living (AAL) program. She wears a bracelet that can detect falls and report Emily's location. Emily is also assisted by Nurse Jane, a professional carer who cares for several other people. Joe is somewhat familiar with computers, so he is able to set up some action patterns which are specific to Emily's case. These are used by the AAL system to help Jane in her activity. One of the main issues in this scenario is to know who is the emergency contact in the case in which a fall is detected.

The next section presents some research related to ours. Existing definitions and previously developed concepts are presented in Section 3, helping the definition of Extended Context Patterns in Section 4. The paper ends with the conclusion and future perspectives.

2   Related Work

There are currently several proposals that intend to deal with the representation of context and situations in a formal manner. Generally they rely on graph theory, and enhance the representation with various types of tags and special kinds of relations.

Bearing a high degree of relation with the formalism of Context Graphs and Patterns, semantic networks [15], concept maps [8] and Sowa's conceptual graphs [14] are directed graphs representing concepts and relations between concepts. One of their main advantages is their property of being graphically displayed, helping understanding of what they express. Conceptual Graphs, in particular, allow expressing any logical formula as a hypergraph. While semantic networks lack some power of expression, which conceptual graphs have, both lack an intuitive mechanism of expressing partially-defined situations. Moreover, these formalisms are not particularly focused on, or appropriate for, the representation of a user's focus and context.

Triples and RDF graphs [7] are easier to use for machines, especially from the point of view of internal representation and disambiguation. However, RDF is difficult to read and write for a user directly, due to the need to use URLs and to repeat the reference to concepts for every relation. Situations recognition can be done by means of SPARQL rules, which however would be difficult to work with by humans. [13]

Some context-focused representations, such as CML (see [12]), use a graphical representation together with a machine-readable XML file which is not adequate for use by humans without assistance from an advanced editor or IDE.

Related to this work are also several methods and tools that have been developed by the authors in previous research. The representation of Context Graphs and Patterns benefits from a visual representation, and also from a basic text representation [9]. More importantly, a matching algorithm has been devised that allows matching context patterns against graphs with very good results for the problem at hand. A platform has also been designed that improves the performance of matching, in the case where the graph evolves over time, and a library of patterns is matched against the same graph.

3   Prerequisites

While there are many definitions of context, we look at two of them in particular. The original definition given by Dey et al was that the *context* contains any element that is relevant to the interaction between the user and the application [4]. Practically, it may extend to the entire current situation of the user. Context elements could in theory be categorized into several categories, such as elements of spatial, temporal, social, or computational context [3], without ignoring activity context [6]. Another definition that is relevant to our approach is the one given by Brézillion et al, comparing context to the dressing of a focus – the context is everything that is related to the current focus of the user [2]. A context-aware application is one that has access to all these elements and uses them in order to provide an improved and more intelligent response to the user.

All the elements that are part of the context need a *representation*. Context-aware behavior is defined by reactions to particular features in the context of the user. Therefore a representation is also needed for the situations that need action to be taken. Such an action may be a notification, a rational inference, or provision of certain information to the user (as does, for instance, the Google Now[1] service). Such representations could be offered by Context Graphs and Context Patterns, respectively.

3.1   Context Graphs and Patterns

We have introduced Context Graphs and a basic version of Context Patterns in previous work [9]. A *Context Graph* is a representation for the current context of

---

[1] https://www.google.com/landing/now/#whatisit

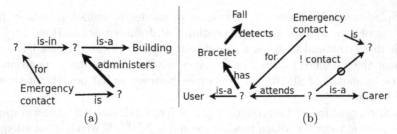

Fig. 1. (a) A Context Pattern showing who should the carer contact in case of an emergency: the administrator of the building where the user is located. The *administers* edge is required for a match to be considered (the edge is *characteristic*). (b) A more complex pattern, containing both *characteristic* (in bold) and *actionable* (marked with a circle) edges. The pattern states that when the bracelet is worn by the user and it detects a fall, the carer who attends the user must be prompted to contact the emergency contact for the user.

the user. It contains a large number of elements, some of which can be categorized as part of the spatial, temporal, computational, social, or activity context. We mainly consider context as a set of elements that are in some *relation* with the user and the user's current situation, therefore any association can be integrated, not only the categories mentioned before.

Formally, the Context Graph of an agent A is a graph CG_A. Considering a global set of *Concepts* (strings or URIs) and a global set of *Relations* (strings, URIs or the empty string λ, for unnamed relations), the graph is defined as:

$CG_A = (V, E)$, where $V \subseteq Concepts$ and

$E = \{ (from, to, value, persistence) \mid from, to \in V, value \in Relations\}$

The *persistence* feature of edges allows for them to expire after a certain time, set when they are created.

In order to detect relevant information, or to find potential problems, an agent has a set of *Context Patterns* that it matches against graph CG_A. These patterns describe situations that are relevant to its activity. A pattern with the identifier s is defined by a graph[2] G_s^P:

$G_s^P = (V_s^P, E_s^P)$, where $V_s^P \subseteq Concepts \cup \{?\}$ and

$E_s^P = \{ (from, to, value, c, a) \mid from, to \in V_s^P, value \in Relations \cup \{\lambda\}\}$

We call nodes labeled with a question mark *generic nodes*. An example of a context pattern is shown in Figure 1(a). It shows that the emergency contact for a user that is in a building is the administrator of that building. Each edge has two features – *characteristic* and *actionable* – showing if a particular edge is absolutely necessary for considering a partial match; and if an edge can be inferred

---

[2] We will use the " ^{P} " superscript to mark structures that support generic elements, such as generic nodes.

(or actioned upon) in case of a partial match[3]. For instance, the *administers* relation is required to exist for the pattern in Figure 1(a) to be considered.

A more complex example of a Context Pattern is shown in Figure 1(b). It contains 2 characteristic edges, that could not be inferred even if the rest of the pattern matches. It also contains an actionable edge: if the rest of the pattern matches with the Context Graph, the Carer will be prompted to contact the person or organization that is resolved to be the emergency contact. How this can be done is shown in the subsequent examples.

By matching a pattern from the agent's set of patterns against the agent's context graph, an agent is able to detect interesting information and is able to decide on appropriate action to take. We have previously developed an efficient algorithm for such matching [9].

The pattern G_s^P *matches* the subgraph $G'_A = (V', E')$, *iff* there exists an injective function $f_v : V_s^P \to V'$, so that the following conditions are met simultaneously:

(1) $\forall v^P \in V_s^P, v^P =?$ or $v^P = f(v^P)$ (same value)

(2a) $\forall (v_i^P, v_j^P, rel) \in E_s^P, (f(v_i^P), f(v_j^P), value) \in E', value \in \{rel, \lambda\}$

(2b) $\forall (v_i^P, v_j^P, \lambda) \in E_s^P, \exists value \in Relations, (f(v_i^P), (v_j^P), value) \in E'$

That is, every non-generic vertex in the pattern has the same label as a different vertex from G'_A (f_v is injective), and every edge in the pattern matches (same label for the edge and vertices) an edge from G'_A. The subgraph G' should be minimal (no edges that are not matched by edges in the pattern). One pattern may match various subgraphs of the context graph. A pattern G_s^P *partially matches with k missing edges (k-matches)* a subgraph G' of G, if conditions (2) above are fulfilled for $m_s - k$ edges in E_s^P, $k \in \{1..m_s - 1\}$, $m_s = \|E_s^P\|$ and G' remains connected and minimal. Partial matches are useful because, depending on a set threshold for k, they indicate cases where action may be taken automatically, to create the missing edges, or notifications may be sent [10].

4 Extended Context Patterns

In order to improve the power of expression held by Context Patterns, the original concept needed to be amended, in order to accommodate the specification of subgraphs with negative character, as well as the operations of alternation and repetition. The result is the *Extended Context Pattern*.

An *Extended Context Pattern* with the name s is defined as
$P_s^E = (H_s^{P_E}, E_s^-, E_s^{(*)}, E_s^|)$,

where $H_s^{P_E} = (V_s^P, E_s^{P_E})$ is the hypergraph underpinning the extended pattern, and E_s^-, $E_s^{(*)}$ and $E_s^|$ are three sets[4] containing information on negative,

---

[3] Both the *characteristic* and the *actionable* features can be numeric instead of boolean, in the interval $[0, 1]$. For simplicity, we will consider them boolean in this work.

[4] the names of the tree sets are read as "E-neg", "E-star", and "E-or".

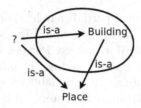

Fig. 2. Example of a hyperedge, represented as an enclosure of a part of the graph. The inbound arity and the outbound arity of the hyperedge are both equal to 1.

repetition and alternation hyperedges, respectively. We will use the notation "P_E" for structures that include hypergraph elements used for extended pattern.

In fact, we only allow $H_s^{P_E}$ some limited structural differences from a normal graph pattern G_s^P. Any edge in the hypergraph is either part of one of the three hyperedge sets, or is a binary, directed edge in the graph pattern.

All three types of hyperedges contained in $H_s^{P_E}$ indicate, in fact, subgraphs of $H_s^{P_E}$ with particular properties. A hyperedge $e_i^{P_E} \in E_s^{P_E}$ *covers* an induced subgraph $H_{si}^{P_E} = (V_{si}^P, E_{si}^{P_E})$, with

$V_{si}^P = e_i^{P_E} \subseteq V_s^P$ and

$E_{si}^P = \{e^{P_E} \mid e^{P_E} \subseteq e_i^{P_E}\}$

It is mandatory that any hyperedge $e_i^{P_E}$ that intersects another hyperedge $e_j^{P_E}$ is either completely included in $e_j^{P_E}$ or completely includes $e_j^{P_E}$, and that no two hyperedges cover the same subgraph. That is,

$\forall\, e_i^{P_E}, e_j^{P_E} \in E_s^{P_E}.\ e_i^{P_E} \cap e_j^{P_E} \neq \emptyset \Rightarrow e_i^{P_E} \subset e_j^{P_E} \vee e_j^{P_E} \subset e_i^{P_E}.$

There may exist, however, some binary, directed edges that have one end inside the graph covered by $e_i^{P_E}$ and one end outside it. We call these edges *arity* edges, and they can be *inbound* or *outbound*. For a hyperedge $e_i^{P_E}$ of the extended graph pattern, the set of arity edges is defined as $\bar{H}_{si}$, with:

$\bar{H}_{si} = \overline{H\text{-}in}_{si} \cup \overline{H\text{-}out}_{si}$

$\overline{H\text{-}in}_{si} = \{e \mid e = (v_k^P, v_l^P) \in E_s^{P_E}, v_k^P \notin e_i^{P_E}, v_l^P \in e_i^{P_E}\}$

$\overline{H\text{-}out}_{si} = \{e \mid e = (v_k^P, v_l^P) \in E_s^{P_E}, v_k^P \in e_i^{P_E}, v_l^P \notin e_i^{P_E}\}$

The *pattern-arity* of a hyperedge $e_i^{P_E}$ is the number of arity edges that it has, that is $\|\bar{H}_{si}\|$. We can define the *inbound pattern-arity* of the hyperedge and its *outbound pattern-arity*.

For instance, Figure 2 shows a graph pattern that contains a hyperedge. The hyperedge covers a graph formed of a single node (*Building*) and no edges, and having one inbound arity edge and one outbound arity edge, amounting to a pattern-arity of 2.

Hyperedges in an extended context pattern do not directly take part in the matching process (i.e. no matching hyperedges are searched for in the context graph), but rather influence how the matching is done.

A **negation hyperedge** $e^\neg \in E_s^\neg \subset E_s^{P_E}, e^\neg \subseteq V_s^P$ covers a subgraph of the context pattern that should not be matched in the context graph, in order to obtain a match of the pattern. Opposite from other edges in the pattern, any

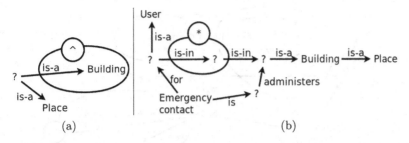

(a) (b)

Fig. 3. (a) An example of a negative hyperedge in a graph pattern specifying a place which is not a building. (b) An example of a pattern containing a repetition hyperedge which may match a longer path of spatial inclusion.

edges that are contained in the graph covered by the negation hyperedge and that are matched with edges in the context graph, increase the k number of the match.

For example, Figure 3(a) shows an example of a negative hyperedge in a graph pattern. The pattern matches a node which is a place but is not a building.

A **repetition hyperedge** $e_i^{(*)} \in E_s^{P_E}$ is part of a tuple

$$(e_i^{(*)}, v_{in}^P, v_{out}^P, e_{in}, e_{out}) \in E_s^{(*)},$$

with $v_{in}^P, v_{out}^P \in e_i^{(*)}$, with e_{in} an arity edge of $e_i^{(*)}$ that is incident to v_{in}^P and e_{out} an arity edge of $e_i^{(*)}$ that is outgoing from v_{out}^P.

The last element of the tuple – e_{out} – is optional. If an *out* edge is specified, the repetition is *binary*, otherwise it is *unary*. A unary repetition hyperedge must have an inbound pattern-arity of at least 1; a binary repetition hyperedge must also have an outbound pattern-arity of at least 1.

In the matching process, the repetition hyperedge acts as a Kleene-star operation on its subgraph. Consider that $e_{in} = (v_a^P, v_{in}^P)$ and, if any, $e_{out} = (v_{out}^P, v_b^P)$. The subgraph covered by a unary repetition hyperedge will match the context graph if:

(1) the context graph contains no subgraph matching H_{si}, or

(2) the context graph (V, E) contains a sequence of n matches of H_{si}, $n \geq 1$, in which v_{in}^P is matched to $v_a \in V$, v_{in}^P is matched to vertices $v_{in}^{(k)} \in V$ and v_{out}^P is matched to vertices $v_{out}^{(k)} \in V$, with $k = \overline{0, n-1}$. Then, there must exist an edge $(v_a, v_{in}^{(0)})$ matching e_{in}, and a series of $n-1$ edges $(v_{out}^{(k)}, v_{in}^{(k+1)})$, $k = \overline{0, n-2}$, also matching e_{in}.

The subgraph covered by a binary repetition hyperedge will match the context graph if:

(1) the context graph contains no subgraph matching H_{si}, but contains an edge (v_a, v_b) matching e_{out}, with v_a matching v_a^P and v_b matching v_b^P; or,

(2) the context graph contains a sequence of n matches of H_{si}, $n \geq 1$, in which v_a^P is matched to $v_a \in V$, v_b^P is matched to $v_b \in V$, v_{in}^P is matched to vertices $v_{in}^{(k)} \in V$ and v_{out}^P is matched to vertices $v_{out}^{(k)} \in V$, with $k = \overline{0, n-1}$. Then, there must exist an edge $(v_a, v_{in}^{(0)})$ matching e_{in}, a series of $n-1$ edges

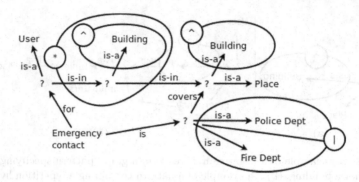

Fig. 4. An example of a pattern containing, among others, an alternation hyperedge.

$(v_{out}^{(k)}, v_{in}^{(k+1)})$, $k = \overline{0, n-2}$, also matching e_{in}, and an edge $(v_{out}^{(n-1)}, v_b)$ matching e_{out}.

For example, Figure 3(b) shows a pattern that serves to determine the emergency contact in the case in which the assisted user is inside a building. The building may have a hierarchy of places (floors, areas, rooms, etc), but only the top node of the hierarchy is a building and has an administrator. This pattern matches any such case.

The formalism may be extended to support the case in which edges between the matches may be different (have a different label) from the edge entering the first match.

An **alternation** operation is characterized by a set of hyperedges with sets of arity edges that are identical from the point of view of direction, label, and adjacent vertex outside of the hyperedge:

$alternation \in E_s^!$ with $alternation \subseteq E_s^{P_E}$, each alternation characterized by two sets:

- $in\text{-}set = \{(v_a^P, label) \mid v_a^P \in V_s^P \setminus \bigcup_{e_i \in alternation} V_{si}^P\}$, the set of sources and labels for arity edges going towards the hyperedge, such that
  $\forall e_i^! \in alternation . \forall (v, u, label) \in \overline{H\text{-}in_{si}} . (v, label) \in in\text{-}set$; and
- $out\text{-}set = \{(label, v_b^P) \mid v_b^P \in V_s^P \setminus \bigcup_{e_i \in alternation} V_{si}^P\}$, the set of destinations and labels for arity edges going towards the hyperedge, such that $\forall e_i^! \in alternation . \forall (v, u, label) \in \overline{H\text{-}out_{si}} . (u, label) \in out\text{-}set$.

The alternation set matches a subgraph of the context graph if the subgraph covered by one of the hyperedges in the alternation correctly matches, as part of the pattern.

The example in Figure 4 builds upon previous examples to show a pattern that helps determine the emergency contact in the case when the assisted user is in an exterior area (not in a building). In this case, the emergency contact is the police or fire department that covers an area which includes the area where the user is located.

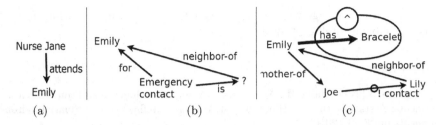

Fig. 5. (a) A simple statement. (b) A simple user-configured pattern. (c) A pattern specific to Emily's case: if Emily does not wear the bracelet, Joe should contact her neighbor Lily.

5   Discussion

Let us look again at the example of Joe and his mother Emily, who is part of an AAL program. Some of the patterns that could be used by a context-aware AAL application have been presented above. They are patterns that show what should happen if a fall is detected, and who is the emergency contact, depending on the context of the user. Such patterns are complex and we cannot expect them to be implemented by normal users. But they are general enough to be already created by developers. Joe can, however, understand what the patterns mean, as the graphical representation is much easier to cope with than other types of representations. Being able to understand why the system chooses to perform an action makes it more acceptable and more dependable.

While a normal user would be unable to create complex patterns, one would surely be able to create simpler ones. For instance, Joe may want to write down that it is Nurse Jane that attends Emily. This can be done by means of a simple graph edge such as the one in Figure 5(a). When Joe understands the system better, he may event insert some patterns, such as the one in Figure 5(b), stating that any of Emily's neighbors can be considered an emergency contact. A more complex user-configured pattern is presented in Figure 5(c). Joe may wish to be prompted to contact Emily's neighbor, Lily, in the case when Emily is not wearing her bracelet (and therefore the AAL system would not have information about Emily's state or whereabouts).

6   Conclusion and Perspectives

This paper presents the visual language of Extended Context Patterns, which builds on the previously developed formalisms for Context Graphs and Context Patterns. Extended Context Patterns allow for the specification of negation, on the one hand, and of variable structures such as repetition and alternatives.

As future work, the matching algorithm must be extended in order to account for the new features of extended context patterns, however the nature of the matching algorithm makes it easy to be adapted to these changes.

Further, the language will be integrated with the tATAmI-2 multi-agent systems for ambient intelligent applications.

References

1. Augusto, J., Nakashima, H., Aghajan, H.: Ambient intelligence and smart environments: A state of the art. Handbook of Ambient Intelligence and Smart Environments pp. 3–31 (2010)
2. Brézillon, J., Brézillon, P.: Context modeling: Context as a dressing of a focus. In: Kokinov, B., Richardson, D., Roth-Berghofer, T., Vieu, L. (eds.) Modeling and Using Context, Lecture Notes in Computer Science, vol. 4635, pp. 136–149. Springer Berlin Heidelberg (2007), http://dx.doi.org/10.1007/978-3-540-74255-5_11
3. Chen, G., Kotz, D.: A survey of context-aware mobile computing research. Technical Report TR2000-381, Dartmouth College (November 2000)
4. Dey, A., Abowd, G., Salber, D.: A context-based infrastructure for smart environments. Proceedings of the 1st International Workshop on Managing Interactions in Smart Environments (MANSE'99) pp. 114–128 (1999)
5. Ducatel, K., Bogdanowicz, M., Scapolo, F., Leijten, J., Burgelman, J.: Scenarios for ambient intelligence in 2010. Tech. rep., Office for Official Publications of the European Communities (February 2001)
6. Henricksen, K., Indulska, J., Rakotonirainy, A.: Modeling context information in pervasive computing systems. Lecture notes in computer science pp. 167–180 (2002), http://www.springerlink.com/content/jbxd2fd5ga045p8w/
7. Lassila, O., Swick, R.: Resource description framework (RDF) model and syntax specification. Tech. rep., The World Wide Web Consortium (1998)
8. Novak, J.D., Cañas, A.J.: The origins of the concept mapping tool and the continuing evolution of the tool. Information Visualization 5(3), 175–184 (2006)
9. Olaru, A.: Context matching for ambient intelligence applications. In: Björner, N., Negru, V., Ida, T., Jebelean, T., Petcu, D., Watt, S., Zaharie, D. (eds.) Proceedings of SYNASC 2013, 15th International Symposium on Symbolic and Numeric Algorithms for Scientific Computing, September 23-26, Timisoara, Romania. pp. 265–272. IEEE CPS (2013)
10. Olaru, A.: Context-awareness in multi-agent systems for ambient intelligence. In: Brézillon, P., Gonzalez, A.J. (eds.) Context in Computing - A Cross-Disciplinary Approach for Modeling the Real World, pp. 541–556. Springer New York (2014), http://link.springer.com/chapter/10.1007/978-1-4939-1887-4_33
11. Perera, C., Zaslavsky, A., Christen, P., Georgakopoulos, D.: Context aware computing for the internet of things: A survey. IEEE Communications Surveys and Tutorials 16(1), 414–454 (2013)
12. Robinson, R., Henricksen, K., Indulska, J.: XCML: A runtime representation for the context modelling language. Pervasive Computing and Communications Workshops, 2007. PerCom Workshops' 07. Fifth Annual IEEE International Conference on pp. 20–26 (2007)
13. Sorici, A., Picard, G., Boissier, O., Florea, A.: Multi-agent based flexible deployment of context management in ambient intelligence applications. In: Advances in Practical Applications of Agents, Multi-Agent Systems, and Sustainability: The PAAMS Collection, pp. 225–239. Springer (2015)
14. Sowa, J.: Conceptual graphs. Foundations of Artificial Intelligence 3, 213–237 (2008)
15. Sowa, J.F.: Semantic networks. Encyclopedia of Cognitive Science (2006)

MDE4IoT: Supporting the Internet of Things with Model-Driven Engineering[*]

Federico Ciccozzi[1] and Romina Spalazzese[2,3]

[1] School of Innovation, Design and Engineering
Mälardalen University, Västerås (Sweden)
federico.ciccozzi@mdh.se,
[2] Department of Computer Science
[3] Internet of Things and People Research Center
Malmö University, Malmö (Sweden)
romina.spalazzese@mah.se

Abstract. The Internet of Things (IoT) unleashes great opportunities to improve our way of living and working through a seamless and highly dynamic cooperation among heterogeneous things including both computer-based systems and physical objects. However, properly dealing with the design, development, deployment and runtime management of IoT applications means to provide solutions for a multitude of challenges related to intelligent distributed systems within the IoT.
In this paper we propose Model-Driven Engineering (MDE) as a key-enabler for applications running on intelligent distributed IoT systems. MDE helps in tackling challenges and supporting the lifecycle of such systems. Specifically, we introduce MDE4IoT, an MDE approach enabling the modelling of things and supporting intelligence as self-adaptation of Emergent Configurations in the IoT. Moreover, we show how MDE, and in particular MDE4IoT, can help in tackling several challenges by providing the Smart Street Lights concrete case.

1 Introduction

Nowadays, connectivity and technology are becoming more and more ubiquitous and affordable. A growing trend is to connect everything that can benefit from being connected, from both digital and physical worlds. Cisco and Ericsson estimated that, by 2020, 50 billions devices will be connected to the Internet [4] and this number is assumed to grow to 500 billions by 2030 with 5G.

The IoT includes a remarkable set of heterogeneous, often distributed, things: sensors, actuators, devices and computers, as well as physical objects that might be equipped with some of the previous elements. An interesting characteristic of IoT systems is that heterogeneity embraces both hardware and software. However, the very same software functionalities are expected to be deployable on different devices having only a limited set of core common features [5].

Another interesting characteristic to observe within the IoT is that things can be lightweight, i.e., with very limited resources and computation capabilities. Within the IoT, we call *Emergent Configuration* (EC) of connected systems

---

[*] This work is partially financed by the Knowledge Foundation through the Internet of Things and People research profile (Malmö University, Sweden) and the SMARTCore project (Mälardalen University, ABB Corporate Research, Ericsson, Alten Sweden).

a set of things/devices with their functionalities and services that connect and cooperate temporarily to achieve a goal. Due to the continuously evolving character of the IoT, ECs can change unpredictably. To be able to provide the user with a coherent IoT system over time, runtime ECs' changes need to be managed. For instance, adaptation mechanisms might re-allocate a set of software functionalities from a faulty device to another offering similar features. In order to maximise reusability of software functionalities (e.g., software components) and minimise the need of functional modifications in response to ECs changes, it would be beneficial to be able to design software functionalities abstracting away platform-specific details, which are instead inferred at deployment time. Finally, to effectively enable collaborative development of IoT systems, mechanisms for supporting separation of concerns are needed.

To benefit from the great advantages that the IoT will unleash, a whole set of challenges related to its intelligent distributed systems needs to be dealt with at all levels. Heterogeneity, adaptability, reusability, interoperability, data mining, security, abstraction, separation of concerns, automation, privacy, middleware and architectures are just some examples of the aspects that need to be taken into account both at design time and at runtime and for which new software engineering approaches shall be envisioned [10, 16, 21, 23]. Among the many challenges, our goal is to provide support for: (1) high-level abstraction to address heterogeneity and system complexity, (2) separation of concerns for collaborative development, (3) intelligence in terms of automated mechanisms for enabling runtime self-adaptation of distributed IoT systems, and (4) reusability.

In this paper we propose MDE as a key-enabler for applications running on intelligent distributed IoT systems. We combine MDE's strengths and the vision of ECs as self-adaptive systems [24] to support design, development, and runtime management of IoT systems. Our novel contributions are: (i) the introduction of MDE4IoT, an MDE approach supporting multiple aspects including modelling, consistency assurance, executables generation at design time, and self-adaptation of ECs at runtime due to evolution. Additionally, we (ii) show how MDE in general, and MDE4IoT in particular, help to tackle the above challenges and to boost self-adaptation within IoT by providing the Smart Street Lights case.

The reminder of the paper is organised as follows. Section 4 introduces the Smart Street Lights case, a distributed intelligent IoT system. In Section 3 we propose the MDE4IoT approach while in Section 5 we describe its application to our case. We present related works in Section 2 and provide discussion and conclusions in Section 6.

2   Relation to the State-of-the-art

The need of exploiting model-driven techniques to help the developer in designing applications for the IoT through separation of concerns and abstraction has been introduced by Patel et al. [23]. They provide automatic generation of an architecture framework and a vocabulary framework to help the developer to manually implement platform-specific portions. On the contrary we aim at supporting the developer by ensuring consistency among design viewpoints and fully generating and deploying executable artefacts, with the developer only focusing on the modelling activities. Conzon et al. [8] focus instead on a lower abstraction level, namely the platform and its software architecture. This approach targets a specific type of IoT configurations, while we aim at providing a generic solutions that can be instantiated in theoretically any type of configuration. Chen et al. [29] present a runtime model-driven approach for IoT application development. While our approach focuses on detailed modelling of software functionalities and their allocation to physical devices, this approach employs models for describing sensor devices only.

Besides few approaches for IoT, the MDE community displays a notable amount of literature addressing adaptation, mostly based on models@runtime. Some approaches use models to specify self-adaptation in terms of mappings of assertions to adaptation actions [18] or to specify links between possible configurations [6]. Approaches that maintain runtime models for specifying adaptation and capturing feedback loop's knowledge exist too [5,9,22]. In [28], Vogel et al. provide an approach that enables the specification and execution of adaptation engines for self-adaptive software with multiple feedback loops.

The most novel characteristic of our approach, which cannot be found in any related work, is the ability to encompass multiple aspects, as modelling, consistency assurance, executables generation at design time, and self-adaptation due to evolution of ECs at runtime. Anyhow, we do not explicitly focus on the use of models for specifying self-adaptation but rather exploit models for the actual adaptation enactment. We plan to focus on how to exploit models for specifying the self-adaptation itself in the coming incarnations of MDE4IoT.

3   A Model-Driven Engineering Approach for IoT

Figure 1 shows a high level model of our vision of IoT self-adaptive systems [24]. Starting from the bottom layer, we find heterogeneous things, which are possibly self-adaptive. Each thing is represented through both its software functionalities and its hardware platform. On the top layer there is a managing system implementing the MAPE-K loop [17]; a relevant aspect is that such a system must have adequate (i) storage space and (ii) computation capabilities to support the management of the system and in particular of the lightweight things.

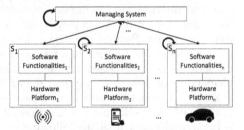

Fig. 1. IoT Self-Adaptive Systems

Why MDE? In MDE, models represent the core concept and are considered an abstraction of the system under development. Rules and constraints for building models are described through a corresponding modelling language definition and, in this respect, a metamodel describes the set of available concepts and wellformedness rules a correct model must conform to [19]. Besides *abstraction*, a core pillar of MDE is the provision of *automation* in terms of model manipulation and refinement, which is performed through model transformations. A model transformation translates a source model to a target model while preserving their wellformedness [7]. Heterogeneity of software and hardware is at the same time a strength and a big challenge within the IoT [23]. Thanks to modelling languages, and more specifically domain-specific modelling languages (DSMLs), MDE can provide unique means for the many aspects of *heterogeneous* systems to be represented all in one place. Models defined through these languages are meant to be much more human-oriented than common code artefacts, which are naturally machine-oriented. This means that, e.g., software can be defined with concepts that are not necessarily dependent on the underlying platform or technology.

Doing so, the very same software functionality should be deployable on heterogeneous physical devices without modifications to enhance *reusability*. Platform-specificity would in fact be inferred by automated mechanisms (i.e., model transformations) in charge of executing models. Translational execution (or code generation), where source models are translated into a third generation programming language (e.g., Java, C++) and then run on the target device after compilation or interpretation, is the one currently provided by MDE4IoT based on our validated code generator [14]. We are already working on possible solutions for providing compilative execution, where source models are directly compiled into executable binary, in MDE4IoT.

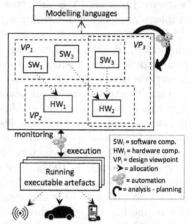

Fig. 2. MDE4IoT Approach

Models can become complex and hard to grasp, even for experts, when heterogeneity is constantly present, thus mechanisms for properly rendering information in ways that are tailored to the specific developer are needed. MDE offers powerful instruments to support *separation of concerns* in terms of multi-view modelling, meant as the ability to define and render models from different design viewpoints. Besides the design and initial deployment of IoT applications, their *evolution at runtime* is a very challenging issue. This is particularly hard if operating on code-based artefacts. Imagine that a specific functionality is implemented for a specific physical device which, at a certain point, stops to be available. It would be hard to re-allocate the functionality to a different type of device without modifying the functionality itself; reusability of the functionality is hence undermined. In the following we describe how we address this.

The MDE4IoT Approach. A graphical representation of MDE4IoT is shown in Figure 2. *Software functionalities* (i.e., deployable software components, SW_i) as well as *physical devices* or *platforms* (i.e., hardware components, HW_i) on which the software functionalities are meant to run, are modelled by means of a set of *modelling languages* (specifically, DSMLs). Deployment of software functionalities to physical devices in terms of *allocations* are modelled too. Physical devices can be represented at different granularity levels. For instance a car navigation system could be represented as a black-box with a specific set of available features and an operating system on which to run the allocated software functionalities. Or it could be divided into smaller pieces, such as sensors, actuators, and processing units. The granularity with which the developer models physical devices depends on the purpose of the models, and the capabilities of the involved model transformations that generate executable artefacts from them. MDE4IoT exploits the combination of DSMLs to achieve *separation of concerns*. In Figure 2 we can see what we call *horizontal* and *vertical viewpoints*. For instance, VP_1 represents a horizontal viewpoint related to a specific software application domain, while VP_2 represents a horizontal viewpoint for physical devices. VP_3

represents a vertical viewpoint since it embraces both software and hardware of a specific application domain. Different viewpoints are concurrently exploited by different domain experts and automated underlying model transformations guarantee consistency among them [3].

Besides assistance at *design time*, MDE4IoT supports *runtime* evolution scenarios within the IoT through self-adaptation mechanisms based on models and model transformations. When an EC changes, and the managing system reasons about possible adaptations, two outcomes are possible: either (1) the affected executable artefacts can be directly re-deployed or (2) the system needs to first re-allocate the functionality at modelling level and then make executable artefacts run on alternative physical devices. Note that (2) represents a typical situation in which human intervention might be needed to drive the re-allocation in case no viable alternatives are available for the managing system to choose among. The managing system's reasoning, although interesting, is out of the scope of this paper (we focus on enabling self-adaptation).

4   The Smart Street Lights Case

Smart cities are one of the application areas of the IoT where everything, including cars, bikes, emergency vehicles, infrastructures and people, will be connected. In this context, we describe an intelligent distributed IoT system: the "Smart Street Lights" demonstrator [26] (basic system) and extend it (scenario). Both its hardware and software have been designed, developed, and assembled through a collaboration between Malmö University and Sigma Technology and is part of the ECOS project [11] within the Internet of Things and People (IoTaP) Research Center [15].

Basic system. The core idea of the Smart Street Lights system is that every car, bike and pedestrian has its own sphere of light provided by a set of smart street lights (or lampposts). The size of this sphere, i.e., the number of lampposts that increase the brightness of their LED lights, is based on the vehicle's or pedestrian's speed and adapts to it at runtime. Yellow lights are dimmed down when nobody is around thus saving energy and red lights are switched on when someone is driving over the speed limit, increasing traffic awareness and safety. The smart lampposts handle all the sensing and computation in a distributed fashion -lampposts are peers running the same code. Each lamppost can: (1) detect the presence of an "object" (car, bike, or pedestrian); (2) compute an object's speed; (3) increase and decrease the brightness of its own lights, either yellow or red; (4) compute the number of lampposts that should increase the brightness of their yellow light or turn on their red lights; (5) send and receive messages to and from neighbour lampposts. Figure 3 illustrates an instance of the Smart Street Lights in use: car A is traveling at normal speed and car B is approaching and traveling over the speed limit thus triggering the red lights to turn on. In the dashed ellipse we can see the set of elements composing a lamppost: a pair of sensors (s_1, s_2), a pair of actuators for yellow and red lights (a_1, a_2), and a computation unit (c_1). A set of lampposts that temporarily connect and cooperate to form the sphere of light accompanying a road user is an *Emergent Configuration*. An interesting *emergent property* of this EC (shown in Figure 3) is that, when a car traveling over the speed limit is approaching another car, bike or pedestrian, the latter gets a heads-up through the red lights.

Scenario. Car A is approaching a street segment where, all of a sudden, the red lights of four subsequent lampposts break (at runtime), and car B is approaching the same street segment while traveling at a speed over the limit. Due to the lampposts malfunction, both car A and car B would not be warned by the system, thus decreasing awareness and safety. To avoid this, the system needs to self-

Fig. 3. Smart Street Lights system

adapt, i.e., to modify itself at runtime to keep its properties. The infrastructures, including the lampposts, are grouped into areas, each having an Area Reference Unit (ARU) providing storage capacity and powerful computation capabilities. The ARU, not shown in Figure 3, takes care of the faulty situation by *initiating (i) a repair procedure and (ii) a system adaptation at runtime*. Initiating the repair procedure for fixing the lights (i) consists of sending a proper message with all the needed information to the service, including for instance the kind of malfunctioning objects and their location. Moreover, (ii) the ARU continuously monitors ECs and detects (a) car A traveling towards the malfunctioning lampposts by exploiting information from the car's navigation system, (b) car B approaching, and, thanks to the lampposts sensors, (c) car B's too high speed. Since the cars' navigation systems are available resources, the ARU (d) sends a warning message (e.g., graphical and/or acoustic) to both cars to reproduce the warning issued by lampposts' red light in normal conditions.

5   MDE4IoT applied to the Smart Street Lights

In this section, we instantiate MDE4IoT to support self-adaptation of ECs in the Smart Street Lights case to concretely highlight the usefulness and advantages of MDE4IoT. Among the many existing modelling languages, UML is regarded as the de facto standard. Its wide adoption is partially motivated by its versatility, enabling its usage as general-purpose language, and customizability through the so-called profiling mechanisms [1] to give it domain specificity through domain-specific profiles. UML's latest incarnation (UML2), together with the standardisation of the Semantics of a Foundational Subset for Executable UML Models (fUML), which gives a precise execution semantics to a subset of UML, and the Action Language for Foundational UML (ALF), to express complex execution behaviours in terms of fUML, have made UML a full-fledged implementation quality language [25]. Through (f)UML and ALF, the developer can fully describe the software functionalities of the system, while exploiting the UML profile for Modeling and Analysis of Real-Time and Embedded Systems (MARTE) [27] for modelling hardware components as well as allocations of software to hardware. Additionally, the Object Constraint Language (OCL) can be used to define specific constraints (e.g., allocation policies) in the model.

Using UML to create domain-specific profiles based on a single metamodel helps MDE4IoT in providing and ensuring consistency among models in multiple viewpoints, in a hybrid multi-view fashion. If the various viewpoints leveraged different domain-specific languages, it would be much harder to ensure consistency since models would conform to different syntactical and semantic definitions. The

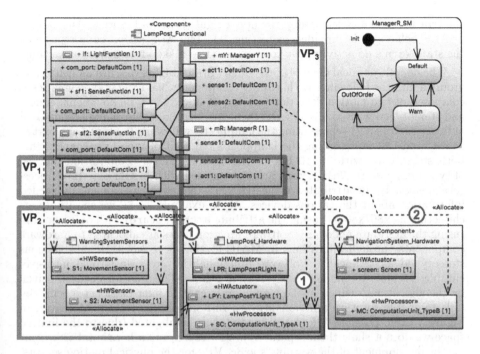

Fig. 4. Smart Street Lights model: ① = basic system, ② = scenario

exploitation of ALF, compliant to UML and platform-independent, instead of third generation programming languages (e.g., Java, C/C++) as action language brings a set of advantages, such as easier model validation, analysis, consistency checking, and the possibility to generate executable artefacts for different targets from the same input models, pivotal for MDE4IoT.

@Design time. In Figure 4 we can see a portion of the model representing the Smart Street Lights in a concrete graphical syntax. More specifically, the portion represents a single lamppost system in terms of software functionalities, physical devices and allocations. In terms of UML, `LampPost_Functional` represents the root software composite component, which contains 6 software components. Among them, `mY` of type `ManagerY` handles sensing and controls the lamppost's yellow light, and `mR` of type `ManagerR` handles sensing and controls the lamppost's red light. Both `mY` and `mR` are connected to `sf1, sf2` of type `SenseFunction` representing the sensing functionality of the two motion detectors. Moreover, `mY` is connected to `lf`, of type `LightFunction`, which represents a lightning functionality (i.e., on/off or dimmed yellow light), while `mR` is connected to `wf`, of type `WarnFunction`, which represents a warning functionality (i.e., on/off of the red light). Connections between software functionalities are achieved through connectors via ports. Behavioural descriptions of the software components are defined in terms of UML state-machines, for defining the overall behaviour by means of states and transitions, and ALF, for providing fine-grained actions. The state-machine diagram describing the behaviour of

```
{
while(!this.toWarn)
if(this.act1.warn() == true)
this.isWarned = true;
}
```

Listing 1.1. Warn's *do activity*

the type ManagerR is depicted in the upper right corner of Figure 4. An example of ALF behaviour is provided in Listing 1.1. The ALF action code represents the state Warn's *do activity*; the code is meant to make wf to display a warning.

The default physical devices of the Smart Street Lights case are modelled and grouped into the following two hardware components. WarningSystemSensors is composed of two motion detection sensors, S1 and S2 of type MovementSensor, and stereotyped with MARTE's «HWSensor». LampPost_Hardware is composed of two actuators, LPY representing the lamppost's yellow light of type LampPostY-Light, and LPR representing the lamppost's red light of type LampPostRLight, both stereotyped with MARTE's «HWActuator», and a computation unit SC of type ComputationUnit_TypeA and stereotyped with MARTE's «HWProcessor». Thanks to the attributes of the MARTE stereotypes we can specify needed information about the various devices. As an example for SC we specify its architecture as x86/x86-64 (using the attribute architecture) with only one core (using the attribute nbCores). Software functionalities are allocated to hardware components through allocations in MARTE, that is to say UML abstractions stereotyped as «Allocate». Software functionalities are transformed into executable artefacts by model transformation chains. The transformations, driven by the allocations defined in the model, are meant to infer platform-specific details to enable software functionalities to run on specific devices. Moreover, in Figure 4 three possible different viewpoints are highlighted (due to the limited space we do not show them separately). VP_1 represents a horizontal viewpoint for the development of the warning's logic, VP_2 for the physical motion sensors, and VP_3 represents a vertical viewpoint of actuators' logic and hardware. Different viewpoints are meant to be concurrently exploited by different domain experts and kept synchronised by underlying mechanisms defined in terms of model transformations [3].

@Runtime. Car A is getting closer to a street segment where suddenly the red lights of four subsequent lampposts break, and car B is approaching the same segment while traveling over the speed limit (scenario). In our model this means that part of LampPost_Hardware suffers from a malfunction. The ARU, our managing system (not shown in the model), initiates a repair procedure by sending a message to the maintenance service, including for instance the kind of malfunctioning objects and their location. Moreover, by continuously monitoring the IoT and ECs, the ARU detects car A and car B as well as their navigation systems as available resources and possible alternatives to replace the malfunctioning red lights. In Figure 4, one of the two navigation systems is depicted. It is represented by NavigationSystem_Hardware, which is composed of an actuator screen, in terms of the navigator's screen, of type Screen and stereotyped with MARTE's «HWActuator», and a computation unit MC of type ComputationUnit_TypeB, stereotyped with MARTE's «HWProcessor» and set as an ARM architecture with two cores. MDE4IoT checks that NavigationSystem_Hardware provides the needed features for hosting the functionalities previously deployed on LampPost_Hardware that should be reallocated: an actuator for showing warnings to replace the broken red light (LPR), and the routines to manage such an actuator to replace part of the intelligence provided by mR. Initially the lamppost's red light manager represented by mR and the warning functionality represented by wf are allocated to LampPost_Hardware's actuator LPR and computation unit SC (① in Figure 4), respectively. MDE4IoT runs a feature-compatibility check[4] between LPR and NavigationSystem_Hardware's actuator represented by screen, as well as between SC and NavigationSystem_Hardware's computation unit represented by MC. If features are compatible, then MDE4IoT re-allocates mR from SC to MC and wf from LPR to screen (② in Figure 4). Re-allocations are meant to be performed automatically by specific in-place model transformations [7] that modify the source models. After a successfull re-allocation, MDE4IoT generates executable artefacts for the newly allocated devices. In case there is no compatibility of features MDE4IoT does not perform any re-allocation, notifies the things about the incompatibility, and awaits for further notifications.

ECs entail unknown devices that must be handled. Things are in charge of providing the minimum information (amount of information can vary) about themselves so that they can be exploited by MDE4IoT to re-allocate the affected software functionality and re-generate suitable executables that can be run on alternative devices. Re-allocation can be driven by an operator when human intelligence is required.

6   Discussion and Conclusion

The Internet of Things (IoT) has a great potential for revolutionising our everyday life in all its aspects. Among other characteristics, the IoT (i) is composed of an unprecedented combination of highly heterogenous constituents, i.e., things, often forming distributed intelligent systems and it (ii) displays continuous evolution, thus leading to the need of methodological innovation.

In this paper we disclosed the opportunities provided by MDE to suffice this need. More specifically, we introduced the MDE4IoT approach to support modelling of things and self-adaptation of Emergent Configurations in the IoT by exploiting: (i) high-level abstraction and separation of concerns to manage heterogeneity and complexity of things, to enable collaborative development, and to enforce reusability of design artefacts, and (ii) automation, in terms of model manipulations, to enable intelligence as runtime self-adaptation.

The incarnation of MDE4IoT presented in this paper represents our first methodological and engineering effort towards an MDE approach for IoT. Individual constituents of MDE4IoT have been employed in more static contexts: multi-view modelling for synchronised separation of concerns in [3], generation of executable artefacts for heterogeneous targets in [14], re-allocation and re-generation based on runtime feedback in [12, 13]. Their synergy, which represents the core of this contribution, had never been investigated before in the IoT community; its validation is ongoing and is meant to leverage a set of industrial use cases. Moreover, the Smart Street Lights case is a validated running prototype [26] of the ECOS project [11].

Self-adaptation of complex, heterogeneous, and variable IoT systems entails an endless set of complex issues to be tackled. The use of models and MDE can help out in this quest, but the way towards a full-fledged solution is far from unhindered. In this paper, we focused on the use of MDE for propagating adaptation from models to the running system. Clearly, other core features need to be investigated as well as future works such as: architectural approaches to

---

[4] Compatibility means that heterogeneous devices are interchangeable if they provide compatible features which can be modelled, e.g., by MARTE's constraints.

self-management [20], the use of models for adaptation, reasoning, and planning, as well as analysis of the possible impact of adaptation-triggered modifications on the unchanged parts of the system [2] .

References

1. A. Abouzahra et al. A practical approach to bridging domain specific languages with UML profiles. In *Procs of OOPSLA*, 2005.
2. A. Bennaceur et al. Mechanisms for leveraging models at runtime in self-adaptive software. In *Models@run.time*. 2014.
3. A. Cicchetti et al. Supporting incremental synchronization in hybrid multi-view modelling. In *Models in Software Engineering*. 2012.
4. K. Ashton. That 'internet of things' thing. *RFiD Journal*, 2009.
5. B. Morin et al. Taming dynamically adaptive systems using models and aspects. In *Procs of ICSE*, 2009.
6. N. Bencomo and G. Blair. Using architecture models to support the generation and operation of component-based adaptive systems. In *Software engineering for self-adaptive systems*. 2009.
7. K. Czarnecki and S. Helsen. Classification of Model Transformation Approaches. In *Procs of OOPSLA*, 2003.
8. D. Conzon et al. Industrial application development exploiting IoT vision and model driven programming. In *Procs of ICIN*, 2015.
9. D. Garlan et al. Rainbow: Architecture-based self-adaptation with reusable infrastructure. *Computer*, 2004.
10. D. Miorandi et al. Internet of things. *Ad Hoc Netw.*, 2012.
11. Emergent Configurations of Connected Systems (ECOS). http://iotap.mah.se/ecos/, [Accessed: 2016-05-19].
12. F. Ciccozzi et al. Round-Trip Support for Extra-functional Property Management in Model-Driven Engineering of Embedded Systems. *Information and Software Technology*, 2012.
13. F. Ciccozzi et al. An Automated Round-trip Support Towards Deployment Assessment in Component-based Embedded Systems. In *Procs of CBSE*. ACM, 2013.
14. F. Ciccozzi et al. On the Generation of Full-fledged Code from UML Profiles and ALF for Complex Systems. In *Procs of ITNG*, 2015.
15. Internet of Things and People (IoTaP) Research Center. http://iotap.mah.se/, [Accessed: 2016-05-19].
16. J. Gubbi et al. Internet of things (iot): A vision, architectural elements, and future directions. *Future Gener. Comput. Syst.*, 2013.
17. J. O. Kephart et al. The vision of autonomic computing. *Computer*, 2003.
18. J. White et al. Simplifying autonomic enterprise Java Bean applications via model-driven engineering and simulation. *Software & Systems Modeling*, 2008.
19. S. Kent. Model Driven Engineering. In *Procs of iFM*.
20. J. Kramer and J. Magee. Self-managed systems: An architectural challenge. In *FOSE '07*, pages 259–268, Washington, DC, USA, 2007. IEEE Computer Society.
21. L. Atzori et al. The internet of things: A survey. *Comput. Netw.*, 2010.
22. M. Amoui et al. Achieving dynamic adaptation via management and interpretation of runtime models. *Journal of Systems and Software*, 2012.
23. P. Patel et al. Enabling high-level application development for the Internet of Things. *Journal of Systems and Software*, 2015.
24. R. de Lemos et al. Software engineering for self-adaptive systems: A second research roadmap. In *Software Engineering for Self-Adaptive Systems II*. 2013.
25. B. Selic. The Less Well Known UML. In *Formal Methods for Model-Driven Engineering*. 2012.
26. The Smart Street Lights Demonstrator. https://vimeo.com/137837738/, [Accessed: 2016-05-19].
27. The UML Profile for MARTE: Modeling and Analysis of Real-Time and Embedded Systems. http://www.omgmarte.org/, [Accessed: 2016-05-29].
28. T. Vogel and H. Giese. Model-driven engineering of self-adaptive software with EUREMA. *ACM Transactions on Autonomous and Adaptive Systems*, 2014.
29. X. Chen et al. Runtime model based approach to IoT application development. *Frontiers of Computer Science*, 2015.

Part III

Security

Detection of traffic anomalies in multi-service networks based on a fuzzy logical inference

Igor Saenko[1], Sergey Ageev[2], and Igor Kotenko[1]

St. Petersburg Institute for Informatics and Automation
of the Russian Academy of Sciences (SPIIRAS),
14-th Liniya, 39, Saint-Petersburg, 199178, Russia

[1]{ibsaen,ivkote}@comsec.spb.ru, [2] serg123_61@mail.ru

Abstract. Methods and algorithms for detection of traffic anomalies in multi-service networks play a key role in creating the malware intrusion detection and prevention systems in modern communication infrastructures. The major requirement imposed to such systems is the ability to find anomalies and, respectively, intrusions in real time. Complexity of this problem is caused in many ways by incompleteness, discrepancy and variety of distribution laws at streams in a multi-service traffic. The paper represents a new technique for traffic anomaly detection in multiservice networks. It is based on using modified adaptation algorithms without identification and fuzzy logical inference rules. Results of an experimental assessment of the technique are discussed.

Keywords: multiservice networks, traffic anomalies, detection and prevention of invasions, stochastic approximation, pseudo-gradient algorithm, fuzzy logical inference.

1   Introduction

Nowadays, the high-speed telecommunication and next generation network technologies have a successful implementation in various information and communication infrastructures (control systems, mobile systems, transport, power supplement, economy, etc.). Progress in development of these technologies has led to the concept of a multi-service network (MSN), which kernel are the basic IP networks integrating services of speech, data and multimedia transmission, and realizing the principle of convergence of telecommunication services [1-3].

The set of main services provided to users by MSN is well known [1-3]. However, emergence of a large number of additional services in MSN make rather sharp the problem of ensuring its information security. This problem increases by following reasons: realization of procedures of dynamic change of the MSN topology; addition or exception of various, a priori uncertain, numbers of network subscribers; dynamic change of spatial arrangement of subscribers; interacting and interfacing MSN with each other, etc. Therefore, the response efficiency of the MSN control system under external and internal destructive influences is of particular importance for the network

© Springer International Publishing AG 2017   79
C. Badica et al. (eds.), *Intelligent Distributed Computing X*,
Studies in Computational Intelligence 678, DOI 10.1007/978-3-319-48829-5_8

security. The normal behavior of network traffic is one of the criteria of safe functioning MSN. At the same time, it is considered that the normal traffic corresponds to a network security policy.

The traffic in the MSN is rather various [2]. It consists of multimedia traffic that is very sensitive to delays, data transmission traffic, alarm information transmission traffic, e-mail traffic, etc. At the same time, the given requirements to the quality of services (QoS) have to be fulfilled completely.

However, there are objective difficulties in creating a MSN management system and protecting the network and subscriber information. These difficulties are caused by complexity of MSN structures, MSN heterogeneity, need to analyze a large number of various network and information parameters. Therefore, fast detection of network traffic anomalies is one of the key problems of the MSN management and represents an actual scientific problem.

The paper presents a novel approach to traffic anomaly detection in a multi-service network based on fuzzy logical influence. At the same time, the rules of fuzzy logical influence are used together with the modified adaptation algorithms without identification and allow to increase stability and convergence of parameters of algorithms. The main theoretical contribution of this paper consists in the following. First, the models for the description of a multi-service traffic are offered. Secondly, the technique of traffic anomaly detection in MSN is developed. At last, experimental confirmation that the developed technique possesses almost greatest possible speed is received. The further structure of the paper is as follows. In section 2 the review of related work is given. The description of the technique of traffic anomaly detection in MSN is provided in section 3. In section 4, experimental results are discussed. Conclusions about the received results and the directions of future research are presented in section 5.

2   Related work

The main architectural decisions for MSN are considered in many works, for example, in [1-3]. These works emphasize that an important feature of the MSN structure consists in a strong segmentation of the network topology and in the existence of several points for interfacing of one MSN with other networks. As a result, the general traffic in the MSN cannot be controlled from one network point.

The technology of intelligent agents for traffic control in networks similar to MSN is proposed in [4, 5]. Intelligent agents carry out network traffic data collection, preliminary processing and transmission of data into the central devices of the control system. In such way, intelligent agents independently develop and realize the part of control functions.

[6] notes the high importance of a problem to ensure that the traffic anomaly detection systems may operate nearly in real time. This work suggests using mixed centralized-decentralized structures for implementation of the MSN control systems. Such structures allow to increase considerably the efficiency of decisions making on

destructive impact counteractions in networks, as well as to decrease the management traffic.

The idea of using intelligent agents for MSN traffic control has gained further development in [7]. This work claims that the implementation of intelligent agents in MSN is impossible without developing and applying algorithms of multi-service traffic anomaly detection, which are simple in realization and robust to change of traffic parameters. At the same time, these algorithms have to operate in real or near real time mode. However, the algorithms offered in this work did not consider fuzzy factors.

Thus, the decisions on algorithms of the network traffic anomaly detection reported in the research literature do not meet the MSN requirements. Mainly, as we think, it is related with the need to process fuzzy information for these algorithms. At the same time, a number of works in which separate attempts to apply methods of processing of fuzzy knowledge are made for modeling and assessment of the network traffic, for example [8-11], is known. However, direct application of the results received in these works for MSN is impossible. This can be explained from the fact that statistical properties of the traffic in MSN strongly differ both for various service conditions, and for various applications in MSN.

There are different techniques to network abnormal traffic detection. For example, the approaches using non-parametric cumulative sums [12], maximum entropy assessment [13], the history of network traffic changes [14], time series from the management data base [15] are proposed. However, these approaches cannot be referred to the methods and algorithms functioning with the given quality in real time mode, and they are not simple enough to implement in MSN.

3   Technique used for traffic anomaly detection

3.1   Multi-service traffic model

To develop a technique of traffic anomaly detection in MSN it is necessary, first of all, to create a multi-service traffic model.

The model of a multi-service traffic in a general view represents an association of a set of various stochastic processes (SP). Therefore, the offered approach to formation of the multi-service traffic model is based on accounting of the following factors: (a) distribution laws for SP; (b) stationarity or non-stationarity of SP; (c) self-similarity of SP; (d) the characteristics of SP chosen for the analysis.

It is noted in [8, 9] that the traffic for various applications in MSN can be approximated by means of various probabilistic distributions. The main distributions are: Poisson, Pareto, Weibull, Log-normal, and Exponential. Various distribution laws are applied to model various traffic types. For example, if the modeled traffic is "Voice" or "Video" then Pareto distributions are applied. If the modeled traffic is created by the SMTP/TCP protocols then Poisson and Exponential distributions are used. The full list of distribution laws for the MSN traffic and their relation with levels of the ISO/OSI model can be found in [16].

Stationarity or non-stationarity of a stochastic process is also an important factor. It is simpler to solve a problem of traffic anomaly detection if it is stationary. However, as it is noted in many works (for example, in [8, 9]), the traffic in MSN is non-stationary by nature. It significantly complicates traffic anomaly detection, as anomalies can be perceived as a normal behavior of the traffic.

Many applications processing a multi-service traffic, for example, audio and video, condition it with self-similarity effect [8, 9]. In this case, stochastic processes have to be described by means of Pareto distribution law.

The traffic analysis can be fulfilled by means of various characteristics of the stochastic process. The main such characteristics are: maximum, minimum and average value of the process intensity, average square deviation and others. In the paper the average value of the process intensity will be considered, which is calculated as follows

$$S_{mid} = \frac{1}{N} \sum_{i=1}^{N} S_i \, , \qquad (1)$$

where N – a size of the selection; S_i – intensity of i-th element in the selection.

Expressions for calculation of other characteristics of the stochastic processes are given in [16].

3.2   Traffic anomaly detection algorithms

The traffic anomaly detection algorithms for MSN have to meet the following requirements: (a) to operate in real or near to real time, (b) to supply given QoS, (c) to have simple realization.

The offered algorithms will belong to the class of hybrid adaptive algorithms of traffic parameters identification. They are applied both for stationary, and non-stationary traffic. Traffic is modeled by stochastic processes (SP). Each SP belongs to the corresponding class.

The essence of the hybrid approach to synthesis of traffic anomaly detection algorithms consists in the following: firstly, the adaptation algorithms without identification [10] are used to the changing SP parameters, and, secondly, the artificial intelligence methods are used for adjustment of algorithm parameters and for decision-making.

Using the hybrid approach is determined by the need to evaluate the current individual and integrated parameters of a traffic. The current integrated parameters of the traffic are evaluated in the "sliding window". The current individual parameters are evaluated sequentially from one point to another point and simultaneously to the integrated assessment procedure. At the same time, the estimates received in the sliding window are used as initial data for adaptation algorithms. For carrying out this procedure two algorithms, differing in the opportunities for approximation of SP, are used.

Combining such approaches allows to estimate integrated SP properties and to trace dynamics of SP behavior. Using methods of logical inference allows to find parametrical estimates of the algorithm parameters based on a small number of evidences. Besides, in this case it becomes possible to draw fuzzy conclusions concerning traffic anomalies.

Let the stochastic process be given in the discrete time points $t_i = i$; $i = 1, 2, \ldots$, that corresponds to real situations of processing on computing tools.

The first algorithm is a modified algorithm of stochastic approximation (MSA) [17]:

$$S_{i+1} = S_{i-1} + \mu_i (S_i - S_{i-1}),\tag{2}$$

where S_i – average value of traffic intensity in the "sliding window" of dimension N at the time point i; μ_i – algorithm parameter at the i-th time point.

The second algorithm is based on application of pseudo-gradient procedures (PGP) [18, 19]:

$$S_{i+1} = S_{i-1} + \mu_i \cdot \mathrm{sign}(S_i - S_{i-1}).\tag{3}$$

Such choice of the first and second algorithms provides an assessment of dynamic SP properties. Because MSA and PGP algorithms belong to the class of the adaptation algorithms without identification, the time of traffic analysis and anomalies detection is significantly reduced.

The square function is offered as a criterion function:

$$M\left[(S_i - S_{i-1})^2\right],\tag{4}$$

where $M[*]$ – the function for a calculation of mathematical expectation value.

Parameter μ for MSA and PGP has to meet the following conditions:

$$0 < \mu < 1, \mu = \mathrm{const.}\tag{5}$$

The peculiarity of MSA and PGP algorithms is the fact that it is necessary to adjust the value of parameter μ for various estimated SPs and their statistical properties.

We propose to carry out the adjustment procedure for these algorithms on a basis of the Mamdani fuzzy inference [5, 10].

Identification of parameters of MSA and PGP algorithms is made by fuzzy logical inference rules. These rules are as follows:

$$\textbf{IF } <S_i = A> \textbf{ AND } <\sigma = B> \textbf{ AND } <\rho = D> \textbf{ THEN } \mu = R,\tag{6}$$

where A, B, and D – fuzzy threshold values defined during training of fuzzy logical inference system [5, 10]. Training is provided on previously created experimental data with a priori known statistical parameters.

Anomaly identification is carried out by fuzzy logical inference rules of the following type:

$$\textbf{IF } <S_i = A> \textbf{ AND } <A \text{ corresponds to the security policy}> \textbf{ THEN } R,\tag{7}$$

where R is a value of the linguistic variable estimating existence of anomalies.

For carrying out the procedure of anomaly identification, it is necessary to train previously the fuzzy logical inference system on experimental data sets as well.

At the choice of the "sliding window" size, it is necessary to find a reasonable compromise between the speed of the change of SP values, the window size and the representativity of the SP value selection. This compromise is necessary for elimination of the effect of excessive smoothing of SP values.

Conclusion about the laws to which the MSN traffic behaviour submits is also carried out by the Mamdani method.

4   Experimental results

An instrumental workbench has been developed for the experimental evaluation of the proposed methods and algorithms (Fig. 1). The workbench consists of the server and two workstations united in the network by a router. One workstation operates as a traffic generator. The administrator works at the second station and controls the generator. The network generator forms a traffic with the given distribution law. At the same time, anomalies are introduced into the traffic in a random way according to the chosen type of a computer attack. The traffic passes through the router on the server. The goal of the administrator is in evaluating and detecting traffic anomalies.

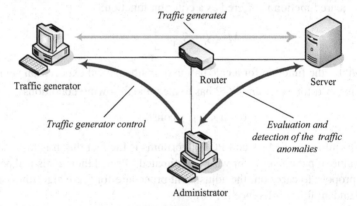

Fig. 1. Instrumental workbench for experimental evaluations

Workbench's software is based on MATHLAB R2013a and implements all techniques described above in Section 3. The generator generates stationary and non-stationary traffic with Poisson and Pareto distributions. These distributions are the most characteristic for MSN. For generation of anomalies the attacks like Denial of Service (DoS) leading to communication channel overload are used. Non-stationary processes are modeled as multiplicative SPs with the determined and stochastic modulating functions. The stochastic modulating functions represent SPs of autoregression (AR) of the first order with various correlation coefficients. Adaptation to the Poisson SP parameters is carried out by means of the MSA algorithm. The PGP algorithm is applied to SP with Pareto law.

Tuning of adaptation algorithm parameters is carried out by means of the rules corresponding (6). The fuzzy constants A, B, D and g for these rules are determined previously by results of training. For this purpose the experimental results presented in Fig. 2 were used.

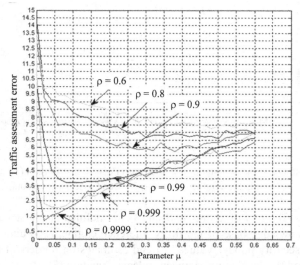

Fig. 2. Dependencies of the traffic assessment error on the parameter μ

Fig. 2 shows dependencies of mean square value of the estimation error of the traffic intensity with non-stationary SP on the value of the MSA algorithm coefficient at various values of a correlation coefficient of the Poisson's SP.

The stochastic process has an average value $a_0 = 100$ and a mean square deviation $\sigma = 10$. The trend correlation coefficient for various dependencies changes in the range from 0.6 to 0.9999. It is visible that at small μ the values of the traffic assessment error strongly differ from each other depending on value ρ. At great μ, these errors accept approximately equal values.

Using the data shown in Fig. 2, rules of control of the MSA parameters are formed. For example, at $S_i \approx 100$, $\sigma \approx 10$ and $\rho \approx 0.9999$ the algorithm parameter μ is approximately equal to 0.025. For $\rho \approx 0.99$ the parameter μ will have the value approximately equal 0.12. The σ and ρ parameters are estimated in the sliding window.

The results of numerical modeling of non-stationary Poisson SP are presented in Fig. 3. The correlation coefficient of AR processes in experiments changes from 0.9 to 0.99999. The SP has the following parameters: initial average value of the trend $m_0 = 500$, correlation coefficient of the trend $r_0 = 0.99$, the "sliding window" size $r = 10$, parameter $\mu = 0.05$, a mean square deviation of the trend $\sigma = 100$. As it can be seen from Fig. 3, the offered algorithm works practically without delays and has rather high estimation precision. In Fig. 3, three zones connected with determination of anomalies are allocated. The zone **A** defines there is the risk of existence of anomalies, and risk level is unacceptable. The zone **B** shows there is the risk of existence of anomalies, and risk level is accepted. In the traffic located in a zone **C**, there are

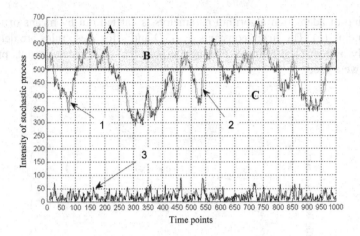

Fig. 3. Assessment of the non-stationary trend with Poisson distribution (1 – true values of the trend; 2 – trend assessments; 3 – trend assessment errors)

no anomalies. Borders of zones **A**, **B** and **C** are defined at the stage of preliminary control of knowledge base of the fuzzy inference engine, according to the accepted rules of the security policy. Values of trend estimates on borders of the zones **A**, **B**, and **C** are used for developing Mamdani rules in equation (7).

Fig. 4 displays results of the assessment of parameters of self-similar traffic with Pareto distribution. Stationary process (Fig. 4-*a*) has the following parameters: the Hurst parameter $H = 0.85$; the average SP value is equal 32.5, the SP distribution parameter is equal 1.3, the MSA algorithm parameter $\mu = 0.004$, the relative estimation error $\Delta \leq 6.2\%$. Non-stationary process (Fig. 4-b) has the following parameters: the Hurst parameter $H = 0.75$; the average SP value is equal 250, the MSA algorithm parameter $\mu = 0.13$, the relative estimation error $\Delta \leq 9.1\%$.

Total results of the assessment of the developed traffic anomalies detection algorithms in MSN for the distributions considered above are given in Tab. 1.

It is visible from the table that the value of the detection error of algorithms does not exceed 12% of true value of a trend. This result should be considered rather good, considering that the algorithms work in real time as the time detection lies in the range from 15 to 80 time points. Thus, the developed algorithms of traffic anomaly detection in MSN completely meet requirements for efficiency and accuracy.

Table 1. Experimental results of an assessment of the algorithms

Distribution low	Correlation coefficient (ρ)	Detection error, %	Detection time, *time points*
Pareto ($H = 0.75$)	0.99	12.0	80
	0.999	10.0	65
	0.9999	8.0	50
Poisson	0.9	8.6	34
	0.99	7.6	27
	0.999	6.4	20
	0.9999	5.3	15

a) stationary trend

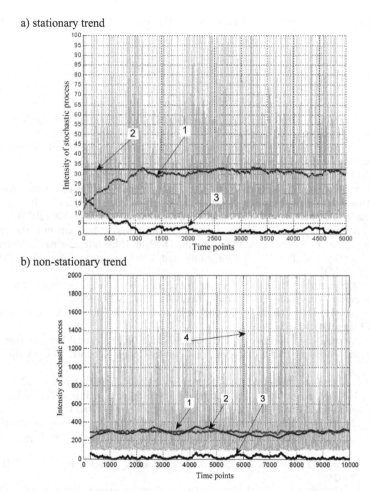

b) non-stationary trend

Fig. 4. Assessment of a trend of the self-similar traffic with Pareto distribution (1 – trend assessment; 2 – true value of the trend; 3 – error of estimation of the trend, 4 – SP values)

5   Conclusions

The paper has proposed a hybrid approach to traffic anomaly detection in MSN using modified adaptation algorithms without identification and Mamdani fuzzy logical inference. Key feature of a multi-service traffic as an object of anomaly estimation is existence of stochastic processes in it subordinated to various distribution laws. For the experimental assessment of the offered algorithms, the Poisson and Pareto distribution laws have been chosen, because they define limiting cases for the regularities of multi-service traffic. The experimental assessment of the offered algorithms has shown that they allow to estimate trends of the stochastic processes in real time with high precision and given QoS. The further directions of researches may be considered as improvement of training methods for proposed algorithms using test traffics.

Acknowledgement. The work is performed by the grant of RSF #15-11-30029 in SPIIRAS.

References

1. Stanwyck, D.: Profitable Deploying Services in the Next Generation Network. In: Voice-in-the Net Japan, Tokyo, Key3Media (2002).
2. Kanáliková, A.: Services In Ngn – Next Generation Networks. In: Journal of Information, Control and Management Systems, Vol. 3, No. 2, pp.97-102 (2005).
3. Wood, R.: Next-Generation Network Services. Cisco Press, USA (2008).
4. Gorodetski, V., Kotenko, I., Karsaev, O.: Multi-agent technologies for computer network security: Attack simulation, intrusion detection and intrusion detection learning. In: International Journal of Computer Systems Science & Engineering, No.4, pp.191-200 (2003).
5. Kotenko, I., Saenko, I., Ageev, S.: Countermeasure Security Risks Management in the Internet of Things based on Fuzzy Logic Inference. In: The 14th IEEE International Conference on Trust, Security and Privacy in Computing and Communications (TrustCom-2015), pp.655-6598 (2015).
6. Paxson, V.: A System for Detecting Network Intruders in Real-Time. In: Proc. of the 7th USENIX Security Symposium, San Antonio, Texas, January 26-29 (1998).
7. Paxson, V.: A System for Detecting Network Intruders in Real-Time. In: Computers Networks, No.31, pp. 2435-2463 (1999).
8. Laskin, N., Lambadaries, I., Harmatzis, F.C., Devetsikiotis, M.: Fractional Levy motion and its application to network traffic modeling. In: Elsevier Comp. Network, Vol. 40, pp.363-375 (2002).
9. Dang, T.D.: New results in multifractal traffic analysis and modeling, Budapest (2002).
10. Ageev, S., Vasil'ev, K.: Adaptive Algorithms for Decorrelation to Image Processing. In: Pattern Recognition and Image Analysis, Vol.11, No.1, pp.131-134 (2001).
11. Takagi, T., Sugeno, M.: Fuzzy Identification of Systems and Its Applications to Modeling and Control. In: IEEE Trans. on System, Man and Cybernetics, Vol.15, No.1, pp.11-132 (1985).
12. Wang, H., Zhang, D., Shin, K.G.: Detecting syn flooding attacks. In: Proc. of IEEE INFOCOM (2002).
13. Staniford, S., Hoagland, J., MCalerney, J.M.: Practical automated detection of stealthy portscans. In: Proc. of the IDS Workshop of the 7th Computer and Communications Security Conference (2000).
14. Brutlag, J.D.: Aberrant behavior detection in time series for network service monitoring. In: Proc. of the 14th Systems Administration Conference, pp.139-146 (2000).
15. Thottan, M., Ji, C.: Anomaly detection in IP networks. In: IEEE Trans. Signal Processing, Vol.51 (2003).
16. Kotenko, I., Saenko, I., Ageev, S., Kopchak, Y.: Abnormal Traffic Detection in networks of the Internet of things based on fuzzy logical inference. In: Proc. of the XVIII International Conference on Soft Computing and Measurements (SCM'2015). IEEE Xplore, pp.5-8 (2015).
17. Spall, L.C. Introduction to Stochastic Search and Optimization: Estimation, Simulation and Control, John Wiley Hoboken, NJ (2003).
18. Polyak, B.T., Tsypkin, Ya.Z. Pseudogradient adaptation and learning algorithms. In: Automation and Remote Control, No. 3(34), pp.377-397 (1973).
19. Calafiore, G., Polyak, B.T., Stochastic algorithms for exact and approximate feasibility of robust LMIs. In: IEEE Trans. on Automatic Control, Vol.46, No.11, pp.1755-1759 (2001).

Reconfiguration of RBAC schemes by genetic algorithms

Igor Saenko and Igor Kotenko

Saint-Petersburg Institute for Informatics and Automation
of the Russian Academy of Sciences (SPIIRAS),
14-th Liniya, 39, Saint-Petersburg, 199178, Russia
{ibsaen,ivkote}@comsec.spb.ru

Abstract. Nowadays, Role-Based Access Control (RBAC) is a widespread access control model. Search of the "users-roles" and "roles-permissions" mappings for the given "users-permissions" mapping is the problem of Data Mining called as Role Mining Problem (RMP). However, in the known works devoted to the RMP, the problem of RBAC scheme reconfiguration and methods for solving it are not considered. The paper defines the statement of the problem of RBAC scheme reconfiguration and suggests a genetic algorithm for solving it. Experimental results show the algorithm has a high enough effectiveness.

Keywords: RBAC, access scheme reconfiguration, genetic algorithms.

1   Introduction

Nowadays, *Role-Based Access Control* (RBAC) is a widespread access control model used in small and big corporative information systems. The main idea of RBAC consists in replacement of the "users-permissions" mapping to consistently connected "users-roles" and "roles-permissions" mappings [1].

Search of the "users-mappings" and "roles-permissions" mappings allowing to receive the "users-permissions" mapping, answering to requirements of a security policy, is a complex challenge from the Data Mining area. This problem named as Role Mining Problem (RMP) [2]. Different variants of the RMP statement are developed, and a lot of methods and algorithms for solving RMP are offered. In all these statements the given "users-permissions" mapping belongs to initial data, and results are the "users-roles" and "roles-permissions" mappings answering to various criteria (for example, to the requirement of the minimum number of roles). Therefore, the known variants of the RMP statement assume that the security policy does not change. However, if there is a need to change the "users-permissions" mapping in corporate system, the security administrator has to solve RMP again. At the same time, the new solution received under solving RMP can strongly differ from previous one. Such solution cannot be always considered as good so as it can involve great administration costs. At the same time, search of the new solution for which RMP criteria cannot be carried out, but administrative expenses will be minimal is of interest. We will call such problem as *reconfiguration of RBAC schemes*. Its main difference from RMP consists in the following: the previous RBAC scheme is introduced in

C. Badica et al. (eds.), *Intelligent Distributed Computing X*,
Studies in Computational Intelligence 678, DOI 10.1007/978-3-319-48829-5_9

the initial data, the criterion is the minimum of the administration expenses. The analysis of the known works devoted to Role Mining shows that methods and algorithms of solving such problems are not known. As RMP, the problem of reconfiguration of RBAC schemes is NP-complete. Development of effective algorithms for solving these problems is of a great interest. A genetic algorithm can successfully apply for a role of such algorithm, as genetic algorithms have well proved for soling many problems in Data Mining [3]. Besides, our previous works [4-6] have shown that genetic algorithms are effective for solving RMP.

The main *theoretical contribution of the paper* consists in the following. First, the problem of reconfiguration of RBAC schemes is formulated for the first time. Secondly, the paper shows that the genetic algorithm can be successfully used for solving this problem, and the paper proposes the main solutions for its realization. *Novelty* of the offered genetic algorithm consists in the following: using two short chromosomes instead of one big one, using the columns of Boolean matrixes as genes of chromosomes, considering the criterion of the minimal administration expenses in the fitness function. The rest of the paper is structured as follows. Section 2 provides an overview of related work. Section 3 deals with mathematical foundations. The description of the genetic algorithm is provided in section 4. Section 5 discusses the experimental results. Conclusions are presented in section 6.

2   Related work

In [2, 7] different variants of RMP statement are formulated. These variants differ in criteria of the "users-roles" and "roles-permissions" mappings under the given "users-permissions" mapping. It is proved that all RMP variants are NP-complete problems.

In [8, 9] the approach based on the cluster analysis is considered. However it demands accounting additional parameters characterizing business processes and needs of users. In [10, 11] the simple heuristic algorithms based on combinatory decisions are suggested. In [12-13] it is offered to reduce complexity of combinatory algorithms due to application of probabilistic models. However this approach does not guarantee high precision of the problem solution. In [14] the approach based on the Boolean Matrix decomposition methods is proposed. In [15] cost-driven technique in which the cost is determined by administration expenses is offered. However this technique did not used to solve the problem of the RBAC scheme reconfiguration. In [16] the metrics to assess various algorithms are considered. Some of these metrics are used in our work. In above mentioned works devoted to solving RMP, the issues of RBAC scheme reconfiguration are not considered. These questions are partially raised in [17-19]. In [17] it is noted that for reconfiguration of RBAC scheme it is possible to use access history logs. In [18] the possibility of reconfiguration of RBAC scheme by means of negative assignments is shown. In [19] statements of RMP with various restrictions are proposed. However they can be considered only as special cases for our problem. Thus, the analysis of relevant work shows that, from one hand, the problem statement of RBAC scheme reconfiguration in them is not considered and, on the other hand, algorithms for solving this problem are not known.

3   Mathematical background

3.1   Traditional RMP statement

Suppose, the RBAC model defines access conditions of m users to n permissions by k roles. Let us introduce the necessary formal notations: $U = \{u_i\}, i = 1,..., m, m = |U|$ – a set of users; $PRMS = \{p_j\}, j = 1,..., n, n = |PRMS|$ – a set of permissions; $ROLES = \{r_l\}, l = 1,..., k, k = |ROLES|$ – a set of roles; $UA \subseteq U \times ROLES$ – a many-to-many mapping of user-to-role assignments; $PA \subseteq PRMS \times ROLES$ – a many-to-many mapping of role-to-permission assignments; $UPA \subseteq U \times PRMS$ – a many-to-many mapping of user-to-permission assignments. The tuple $< U, PRMS, ROLES, UA, PA >$ is named as *RBAC configuration*, or *RBAC scheme*. Further, we will use the second definition. Traditional RMP consists in the following: for given U, $PRMS$, $ROLES$, and UPA to find UA and PA, which would meet some criterion. The type of this criterion defines different RMP variants. The most popular RMP variants are as follows: *Basic* RMP, if the number of roles k should be minimal; *Edge* RMP, if the total number of assignments in UA and PA has to be minimal; *Minimal Noise* RMP, if under the given k the divergence between the given UPA mapping and the "user-permissions" mapping, which turns out by means of the found UA and PA, should be minimal; *δ-Approx* RMP, if the divergence between the given UPA mapping and the "user-permissions" mapping, which turns out by means of the found UA and PA, does not exceed value δ, and k should be minimal. The example of graphical representation of a RBAC scheme is given in Fig.1 [7].

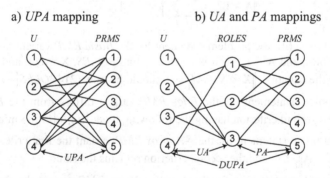

Fig. 1. Example of mappings between sets of users, permissions and roles

Fig.1*a* shows the mapping between U and *PRMS* which is required to be received by means of the RBAC scheme. Fig. 1*b* demonstrates how the mapping between U and *PRMS* is formed, if consistently to connect two mappings UA and PA. The mapping which is turning out as a result of consecutive connection of UA and PA mapping is called as *Direct UPA*, or *DUPA*. The UA and PA mappings need to be found. The UPA mapping is given. It is easy to find out that the example given in Fig.1 corresponds to the *Basic* RMP and *Edge* RMP variants. The *DUPA* mapping, which turns out at the consecutive UA and PA connection, completely coincides with the required UPA mapping. At the same time, value k accepts the minimum value equal 3, and the total number of assignments in the UA and PA mappings is equal to 13.

The UPA mapping corresponds to $m \times n$ Boolean matrix $\mathbf{A} = M(UPA)$, in which "1" in cell $\{i, j\}$ indicates the assignment of permission j to user i. Similarly, we will determine matrixes $\mathbf{X} = M(UA)$ and $\mathbf{Y} = M(PA)$.

In case of *Basic RMP* variant, matrixes $\mathbf{A}$, $\mathbf{X}$, and $\mathbf{Y}$ are as follows:

$$\mathbf{A} = \mathbf{X} \otimes \mathbf{Y}, \tag{1}$$

where symbol $\otimes$ stands for Boolean matrix multiplication, which is a form of matrix multiplication based on the rules of Boolean algebra. Boolean matrix multiplication allows getting the elements of the matrix $\mathbf{A}$ by the following expression: $a_{il} = \vee_{j=1}^{n}(x_{ij} \wedge y_{jl})$, where x_{ij} is the element of matrix $\mathbf{X}$ in the cell $\{i, j\}$, y_{jl} is the element of matrix $\mathbf{Y}$ in cell $\{j, l\}$.

Using expression (1), the example, given in Fig.1, can be written down in the form of the following expression:

$$\begin{pmatrix} 0 & 1 & 0 & 0 & 1 \\ 1 & 1 & 1 & 0 & 1 \\ 1 & 1 & 0 & 1 & 1 \\ 1 & 1 & 1 & 0 & 0 \end{pmatrix} = \begin{pmatrix} 0 & 0 & 1 \\ 1 & 0 & 1 \\ 0 & 1 & 1 \\ 1 & 0 & 0 \end{pmatrix} \otimes \begin{pmatrix} 1 & 1 & 1 & 0 & 0 \\ 1 & 1 & 0 & 1 & 0 \\ 0 & 1 & 0 & 0 & 1 \end{pmatrix}.$$

To define distance between Boolean matrixes, the concept L_1-*norm* is used [7, 17]. For matrixes $\mathbf{A}$ and $\mathbf{B}$ of equal dimensions this metrics is calculated as follows:

$$\|\mathbf{A} - \mathbf{B}\|_1 = \sum_{i=1}^{m} \sum_{j=1}^{n} |a_{ij} - b_{ij}|. \tag{2}$$

Using (1) and (2), the problem statement for the *Basic RMP* variant is as follows. Given U, $PRMS$, and $\mathbf{A} = UPA$, it is needed to find $ROLES$, $\mathbf{X} = UA$, and $\mathbf{Y} = PA$ in order to: (1) the equation $\|\mathbf{X} \otimes \mathbf{Y} - \mathbf{A}\|_1 = 0$ should be true, (2) $|ROLES| \rightarrow \min$.

The problem statement for the *Edge RMP* variant differs from the *Basic RMP* variant by the second criterion having the following type: $|UA| + |PA| \rightarrow \min$.

In the problem statement for the *δ-Approx RMP* variant the first criterion has the type $\|\mathbf{X} \otimes \mathbf{Y} - \mathbf{A}\|_1 \leq \delta$ and the second criterion remains the same.

In the problem statement for the *Minimal Noise RMP* variant the first criterion has the type $\|\mathbf{X} \otimes \mathbf{Y} - \mathbf{A}\|_1 \rightarrow \min$ and the second criterion means $|ROLES|$ is given.

[7] shows that all RMP variants are *NP*-complete problems.

3.2 Essence of the problem of RBAC scheme reconfiguration

The essence of the problem of RBAC scheme reconfiguration consists in the following. Let a security policy has some changes, because of which in the existing UPA mapping one or several assignments change (i.e. they are added or removed). It is necessary to find corresponding changes in the *UA* and *PA* displays which would minimize administration expenses upon transition to the new RBAC scheme.

Consequences from change in the *UPA* mapping of only one assignment can be various. In one case at the same time it will also be required to change one assignment in the *UA* or *PA* mappings. In other case change of a set of roles (by addition of a new role) and change more than one assignments in the *UA* or *PA* mappings can be required. These examples are illustrated below in Fig. 2 and Fig. 3. Dashed lines correspond to deleted assignments, continuous fat lines – to added assignments.

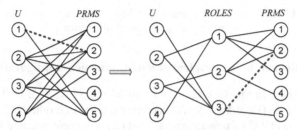

Fig. 2. Example of the change of one assignment in *UPA* causing change of one assignment in *PA*

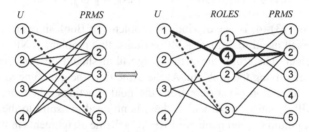

Fig. 3. Example of the change in *UPA* causing multiple changes in *UA* and *PA*

Fig. 2 shows the example, when one assignment between user 1 and permission 2 is removed from the *UPA* mapping. At such change of the security policy it is enough to remove in the RBAC scheme one assignment between the role 3 and the permission 2 in the *PA* mapping. Fig. 3 shows more complex example, when in the *UPA* mapping one assignment – between the user 1 and the permission 5 – is also removed. However at such change of the security policy in the RBAC scheme it is necessary to remove one assignment (between the user 1 and the role 3), to add a new role 4 and to add two assignments connecting the role 4 to the user 1 and the permission 2. In this case in the *UA* and *PA* mappings it is necessary to change 3 assignments. However if the problem given in Fig.3 to solve as the traditional RMP, then the solutions given in Fig. 4 turn out. Fig. 4*a* shows the result of solving the Basic RMP. In this case it is necessary to change 6 assignments. Fig. 4*b* shows the result of solving the Edge RMP. In this case it is necessary to change 9 assignments. Solving the problem of RBAC scheme reconfiguration by traditional RMP methods cannot be considered as the best by criterion of minimal administration expenses. Thus, the formal problem definition of RBAC scheme reconfiguration is necessary.

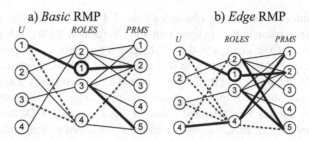

Fig. 4. Example of the solution of the reconfiguration problem
by means of traditional RMP variants

3.3   The statement of the problem of RBAC scheme reconfiguration

The initial data in this task are: (1) an initial matrix $\mathbf{A}_0 = M\,(UPA_0)$; (2) initial matrixes $\mathbf{X}_0 = M\,(UA)$ and $\mathbf{Y}_0 = M\,(PA)$ connected with each other and with matrix A by the expression (1); (3) a new matrix $\mathbf{A}_1 = M\,(UPA_1)$, corresponding to a new security policy. It is required to find such matrixes $\Delta\mathbf{X}$ and $\Delta\mathbf{Y}$ that:

$$(\mathbf{X}_0 \oplus \Delta\mathbf{X}) \otimes (\,\mathbf{Y}_0 \oplus \Delta\mathbf{Y}) = \mathbf{A}_1. \qquad \left\|\Delta\mathbf{X}\right\|_1 + \left\|\Delta\mathbf{Y}\right\|_1 \to \min. \qquad (3)$$

The problem (3)-(4) is one of kinds of problems of Boolean matrix factorization. Therefore, as well as all traditional RMP variants, this problem is NP-complete optimization problem. Therefore, the genetic algorithm should be considered as rather effective way of solving the problem. At the same time the goal function of the problem given in expression (4) differs from the goal functions which are available in traditional RMP variants. In particular, there is no need to aspire to the minimum of roles or the minimum of total number of RBAC scheme assignments in this problem.

4   Genetic algorithm

4.1   Chromosomes and an initial population of individuals

We will use as the possible solution not a matrix $\Delta\mathbf{X}$ which plays the role of the variable in the problem (3), but the matrix $\mathbf{X}_1$, which is defined as the sum (excluding OR) of $\Delta\mathbf{X}$ and the known matrix $\mathbf{X}_0$. Such choice allows to use the earlier developed genetic algorithm discussed in [4-6] for solving RMP and finding $\mathbf{X}_1$.

As we can see from the problem definition (3), matrixes $\Delta\mathbf{X}$ and $\Delta\mathbf{Y}$ play the role of variables. As a rule, in genetic algorithms the chromosomes have to code required variables. Therefore for our problem, chromosomes have to include elements of these matrixes. However, if to build all elements of matrixes $\Delta\mathbf{X}$ and $\Delta\mathbf{Y}$ in one string consisting of Boolean elements, then there is a number of difficulties.

First, such chromosome will have the variable length as in the intermediate RBAC schemes the quantity of roles can be arbitrary. Secondly, the procedure of the adaptive choice of a point of crossing, in which parental chromosomes share for an exchange of the parts, is necessary. Realization of such procedure is rather complex challenge. Thirdly, during the crossing and mutation there will be a high percent of formation of invalid individuals which at once are subject to removal. It will have an adverse effect on convergence of the genetic algorithm.

To avoid these shortcomings, we propose the following solutions. First, each of matrixes ΔX or ΔY will be used for formation of separate chromosomes. Secondly, not elements, but columns of Boolean matrixes will be used as genes of these chromosomes. Lengths of these chromosomes will be limited by the greatest possible number of roles in the RBAC scheme. This number is defined as $K = \min \{m, n\}$.

4.2   Fitness function

We will form the fitness function as follows:

$$F = \left(w_1 F_1 + w_2 F_2\right)^{-1}, \tag{4}$$

where w_1 and w_2 – weight coefficients; F_1 – function which needs to be minimized; F_2 – function which reflects full coincidence of the initial and resultant UPA mappings. The ratio $w_1 \ll w_2$ is established between weight coefficients. This ratio guarantees that during work of the genetic algorithm at the first stage there is a search of solutions, where $F_2 = 0$, and then there is a search of solutions to smaller values F_1.

Considering (3) and (2), it is offered to create the function F_1 as follows:

$$F_1 = \sum_{i=1}^{m}\sum_{l=1}^{K}\left|x0_{il} \oplus x1_{il}\right| + \sum_{j=1}^{n}\sum_{l=1}^{K}\left|y0_{jl} \oplus y1_{jl}\right|, \tag{5}$$

where $\{x0_{il}\}$, $\{x1_{il}\}$, $\{y0_{jl}\}$ and $\{y1_{jl}\}$ – elements of the matrixes X_0, $X_1 = X_0 \oplus \Delta X$, Y_0 and $Y_1 = Y_0 \oplus \Delta Y$ respectively.

Considering (1), (2) and (3), it is offered to create the function F_2 as follows:

$$F_2 = \sum_{i=1}^{n}\sum_{j=1}^{m}\left|a1_{ij} - \sum_{l=1}^{K}x1_{il}\,y1_{lj}\right|, \tag{6}$$

where $\{a1_{ij}\}$ – elements of the matrix A_1.

4.3   Crossing and mutation

Performance of the crossing operation in a case when each individual has two chromosomes, assumes that each of these chromosomes will have own point of crossing. In this case parental individuals will have two exchange channels. Across the first canal there will pass the exchange of parts of the first chromosome, across the second canal – the second chromosome. As a result we will offer that as a result of performance of the crossing operation $4 = 2^2$ various descendants are formed. It will allow to increase significantly the speed of the genetic algorithm work.

As in intermediate problem solutions there can be various numbers of roles, it is necessary to provide a possibility of crossing of the individuals reflecting solutions with various number of roles. For this purpose, it is offered before crossing and after it to carry out additional processing of chromosomes. Before crossing this processing assumes performance of operation of gene sorting during which the genes with zero columns move to the end of the chromosome. After crossing if one of two chromosomes has a zero column, then it is necessary be nullify a column with the same number at other chromosome. Thereby the coherence of both chromosomes is provided. They will display identical number of roles in the *UA* and *PA* mappings.

5   Experimental results

The experimental assessment of the genetic algorithm offered for solving the problem of RBAC scheme reconfiguration was carried out at the software tool which carried out the following functions: (1) Random generation of mappings *UA* и *PA* on the base of the given values m, n and $k = m / 3$. The result is fixed in matrixes $X_0 = M(UA)$ and $Y_0 = M(PA)$; (2) Forming the mapping *DUPA* in consequence with equation (3). The result is fixed in the matrix $A_0 = M(DUPA)$; (3) Random generation of changes ΔA, carrying out to new matrix A_1, connected with the matrix A_0 by equation $A_1 = A_0 \oplus \Delta A$; (4) The work of the genetic algorithm that solves the optimization problem (3)–(4). The result is fixed in matrixes X_1 и Y_1; (5) The work of the genetic algorithm that solves the optimization problem *Basic* RMP. The result is fixed in matrixes X_2 и Y_2 ; (6) Comparative assessment of the found solutions. The distance metrics are used: $L_1 = \|X_1\|_1 + \|Y_1\|_1$ and $L_2 = \|X_2\|_1 + \|Y_2\|_1$.

Generation of mappings was carried out for the following couples of values (m, n): (9, 30), (21, 100) and (36, 500). The choice of these values was defined by the purpose to define how the algorithm speed is changed under increasing the dimension of the problem. The choice of value $k = m / 3$ has been caused by desire to make the generated RBAC scheme more realistic. The number of assignments in the generated scheme was estimated as $L_0 = \|A_0\|_1$. Under generation of changes ΔA the coefficient γ named as *reconfiguration power* was considered. It was defined as L_Δ / L_0, where $L_\Delta = \|\Delta A\|_1$. The coefficient γ had the values 0.1, 0.25, and 0.5. At higher value γ, reconfiguration loses meaning, and solving traditional RMP is necessary. The number of iterations T that are spent for search of X_1 and Y_1 X_1 was considered as the algorithm speed indicator. Results of experiments are given in Table 1.

Values of indicators L_Δ, L_1, L_2 and T are calculated as averages on the random selection making 10 tests. Analyzing the data in Table 1 it is possible to draw the following conclusions. At small dimension of the problem (9, 30) and at different γ the values of functions L_1 and L_2 are approximately equal. It means that the results of solving traditional RMP and the reconfiguration problem are almost identical. The same effect is observed at dimensions (21, 100) or (36, 500), but at small value $\gamma = 0.10$. At big γ (0.25 or 0.50) the result of the reconfiguration has advantage in comparison with solving RMP, and the gain is larger, when the larger (n, m) and γ.

The analysis of the proposed algorithm speed shows that with growth of (m, n) and γ the number of demanded iterations has the dependence that is close to linear.

That is acceptable for the NP-complete task. Therefore, it is possible to draw a conclusion that the proposed genetic algorithm is efficient.

Table 1. Experimental results.

m	n	L_0	γ	L_Δ	L_1	L_2	T
9	30	135	0.10	13.5	22.7	22.1	77.0
			0.25	33.8	55.9	55.0	114.6
			0.50	67.5	113.0	114.8	139.1
21	100	1050	0.10	105.0	180.3	187.3	437.5
			0.25	262.5	416.5	479.5	735.0
			0.50	525.0	787.5	843.5	927.5
36	500	9000	0.10	900	1281	1361	4134
			0.25	2250	3037	3897	5404
			0.50	4500	3759	4647	6240

The received experimental results allow to make a number of recommendations for the RBAC administration. First of all, it should be noted that if RBAC scheme is created for the first time then the administrator has to implement the genetic algorithm to solve RBAC which minimizes the number of roles at full coincidence of mappings *UPA* and *DUPA*. Further actions for the RBAC scheme reconfiguration depend on dimension of the scheme. The dimension of the RBAC scheme is defined by the total number of users and permissions. If the RBAC scheme has low dimension, then at change of the given UPA mapping the solutions of the traditional RMP and the RBAC scheme reconfiguration problem have identical importance. We recommend using solving the traditional RMP in interests of additional optimization of the RBAC scheme. If the RBAC scheme has high dimension then at a change of the *UPA* mapping the administrator is obliged to use the reconfiguration which causes considerably smaller number of changes in the available RBAC scheme, than in the traditional RMP. It is necessary to notice that even if the *UPA* mapping changes for 10-15 percent, the number of necessary changes in the RBAC scheme is so big that the reconfiguration in the manual mode is almost impossible.

6   Conclusion

The paper proposes the approach to the solution of the reconfiguration problem for the RBAC scheme based on application of the developed genetic algorithm. The analysis of the mathematical problem statement has shown that the problem is closely connected with traditional RMP variants and is NP-complete. It differs from traditional RMP variants in structure of initial data (the existing RBAC scheme concerns to them) and optimization criterion which consists in minimization of the administration costs. The genetic algorithm is developed to solve the problem. Distinctive features of the algorithm are: presence of two chromosomes which genes are columns of "users-roles" and "roles-permissions" matrixes at each individual; additional processing of chromosomes before and after crossing; considering criterion of the minimum administrative cost in the fitness function. The experimental assessment of the

98

I. Saenko and I. Kotenko

proposed genetic algorithm showed its sufficiently high efficiency. Future research is related with the analysis of the approach for the RBAC schemes of high dimension.

Acknowledgement. The work is performed by the grant of RSF #15-11-30029 in SPIIRAS.

References

1. Sandhu, R.S., Coyne, E.J., Feinstein, H.L., Youman, Ch.E.: Role-Based Access Control Models. In: Computer, vol. 29 n.2, pp.38-47 (1996).
2. Frank, M., Buhmann, J.M., Basin, D.: On the Definition of Role Mining. In: 15th ACM symposium on Access control models and technologies, pp. 35-44 (2010).
3. Verma, G., Verma, V.: Role and Applications of Genetic Algorithm in Data Mining. In: International Journal of Computer Applications, vol. 48, is.17, pp. 5-8 (2012).
4. Saenko, I., Kotenko, I.: Genetic Algorithms for Role Mining Problem. In: PDP 2011 Conference, pp. 646-650 (2011).
5. Saenko, I., Kotenko, I.: Design and Performance Evaluation of Improved Genetic Algorithm for Role Mining Problem. In: PDP 2011 Conference, pp. 269-274 (2012).
6. Kotenko, I., Saenko, I.: Improved genetic algorithms for solving the optimization tasks in access scheme design for computer networks. In: International Journal of Bio-Inspired Computation, vol.7, no.2, pp.98-110 (2015).
7. Vaidya, J., Atluri, V., Guo, Q.: The Role Mining Problem: Finding a Minimal Descriptive Set of Roles. In: ACM symposium on Access control models, pp.175-184 (2007).
8. Kuhlmann, M., Shohat, D., Schimpf, G.: Role Mining – Revealing Business Roles for Security Administration using Data Mining Technology. In: 8th ACM symposium on Access control models and technologies, pp.179-186 (2003).
9. Lu, H., Hong, Y., Yang, Y., Duan, L., Badar, N.: Towards User-Oriented RBAC Model. In: Journal of Computer Security, vol.23, is.1, pp. 107-129 (2015).
10. Vaidya, J., Atluri, V., Warner, J.: RoleMiner: mining roles using subset enumeration. In: 13th ACM conference on Computer and communications security, pp. 144-153 (2006).
11. Blundo, C., Cimato, S.: A Simple Role Mining Algorithm. In: 2010 ACM Symposium on Applied Computing, pp. 1958-1962 (2010).
12. Colantonio, A., Di Pietro, R., Ocello, A., Verde, N.V.: A Probabilistic Bound on the Basic Role Mining Problem and its Applications. In: IFIP Advances in Information and Communication Technology, vol.297, pp. 376-386 (2009).
13. Frank, M., Buhmann, J.M., Basin, D.: Role Mining with Probabilistic Models. In: ACM Transactions on Information and System Security, vol.15, is.4, article no.15 (2013).
14. Frank, M., Streich, A.P., Basin, D.: Multi-Assignment Clustering for Boolean Data. In: The Journal of Machine Learning Research, Vol.13, no.1, pp. 459-489 (2012).
15. Colantonio, A., Di Pietro, R., Ocello: A cost-driven approach to role engineering. In: 2008 ACM symposium on Applied computing, pp. 2129-2136 (2008).
16. Molloy, I., Li, N., Li, T. Lobo, J.: Evaluating Role Mining Algorithms. In: 14th ACM symposium on Access control models and technologies, pp. 95-104 (2009).
17. Jafari, M., Chinaei, A.H., Barker, K., Fathian, M.: Role Mining in Access History Logs. In: International Journal of Computer Information Systems and Industrial Management Applications, vol.1, no.1, pp. 258-265 (2009).
18. Uzun, E., Atluri, V., Lu, H., Vaidya, J.: An Optimization Model for the Extended Role Mining Problem. In: LNCS, vol.6818, pp. 76-89 (2011).
19. Blundo, C., Cimato, S.: Constrained Role Mining. In: LNCS, vol.7783, pp. 289-304 (2013).

String-based Malware Detection for Android Environments

Alejandro Martín[1], Héctor D. Menéndez[2], and David Camacho[1]

[1] Universidad Autónoma de Madrid, Madrid 28049, Spain,
{alejandro.martin,david.camacho}@uam.es,
[2] University College London, London WC1E 6BT, UK
h.menendez@ucl.ac.uk,

Abstract. Android platforms are known as the less security smartphone devices. The increasing number of malicious apps published on Android markets suppose an important threat to users sensitive data, compromising more devices everyday. The commercial solutions that aims to fight against this malware are based on signature methodologies whose detection ratio is low. Furthermore, these engines can be easily defeated by obfuscation techniques, which are extremely common in app plagiarism. This work aims to improve malware detection using only the binary information and the permissions that are normally used by the anti-virus engines, in order to provide a scalable solution based on machine learning. In order to evaluate the performance of this approach, we carry out our experiments using 5000 malware and 5000 benign-ware, and compare the results with 56 Anti-Virus Engines from VirusTotal.

Keywords: Malware, Classification, Android

1 Introduction

Android is currently one of the most relevant Operative Systems. The amount of software which is published on different Android stores makes this systems one of the most profitable for developers, too.

Detecting malware is a costly procedure, since the concealment sophistication has produced strong bottlenecks, following that both, static and dynamic analysis techniques, become almost useless during the detection process. This has been specially problematic with Windows malware [8], and has also become a problem in new technologies malware, such as Android [19].

One of the most promising detection methodologies are those based on machine learning [21], which are usually supporting static and dynamic approaches. Machine learning is divided in different fields [10], where the most relevant for this work are classification and clustering [11]. These techniques has been successfully applied in several different and heterogeneous areas, such as: social networks [2], autonomous systems [16], video-games [17] and optimization [12]. The scalability of these techniques make them really useful for Big Data [3] problems as is the case of detecting malware in a general way.

© Springer International Publishing AG 2017   99
C. Badica et al. (eds.), *Intelligent Distributed Computing X*,
Studies in Computational Intelligence 678, DOI 10.1007/978-3-319-48829-5_10

This work aims to find a trustful malware detection methodology based on classification using only the binary and permissions information that can be extracted from the apk file. This methodology, known as string based detection, is a scalable solution due to it can deal with the different malware without any disassembly or running process.

The new methodology aims to improve the detection using supervised learning. This combination based on binary structure learning, supposes an important advantage compared with similar solutions such as signature-based detection, usually used by commercial Anti-Virus.

In order to evaluate the performance of the new methodology, we have download a corpus of 5000 malware and 5000 benign-ware from different public sources. Besides, in order to show the importance of our technique, we have compared the accuracy results with the 56 Anti-Virus Engines available in VirusTotal. The results have shown that our methodology obtains a strong discrimination accuracy (more than 0.97) and it overcomes the best Anti-Virus Engine in more than 10 points according to the detection accuracy (0.97 versus 0.83 accuracy obtain by the best Anti-Virus).

The paper is structured as follows, next section presents a short related work introducing malware detection methodologies, specially focused on Android detection techniques. After, we present our modelling process for detecting Android malware using classification. Section 4 shows the main research goals and evaluation conditions that are evaluated in Section 5. Finally, the last section shows the conclusions and future work.

2   Related Work

Malware detection is a sensitive task and maybe the most important step in order to prevent Malware. There are several works that have been focused on this problem. A good review of some of them can be found in [8], where Idika and Mathur present a survey about 45 Malware detection techniques. The authors categorise these techniques according to three main approaches:

- **Static Analysis**: it is focused on analysing malware before running it, using the program control flow. It is usually performed using a disassembly process.
- **Dynamic Analysis**: this analysis is based on malware execution behaviour. It is focused on system and network traces, memory, and process execution, among other features.
- **Hybrid Analysis**: this methodology combines the two previous methodologies in order to improve the accuracy during the detection process.

Static analysis is usually focused on disassembling the code and studying the assembly version generated. Some works are focused on the operational codes (or opcodes) of this assembly code. For example, Santos et al. [18] model their detector using frequencies of opcodes and providing a sequence probability to them, these opcodes are obtained from malware disassembly code. The opcode frequency and the mutual information are combined to measure the similarity of two program behaviours. Other authors use semantic representations of the code in order to understand its behaviour for the discrimination, such as Preda et al. [14] or Christodorescu et al. [4]. It is important to remark that static analysis has limitations, as Moser et al. studied in [13], where they were focused on finding the limits of static approaches. Authors present the main problems of detection techniques based on static analysis, specially signature-based techniques which are avoided using obfuscation. They also introduce the trade-off of semantic techniques analysing its effectiveness with zero-day Malware.

Dynamic analysis is featured by taking into account the execution events. Good examples can be found applied to new technologies such as Android Malware analysis. For example, Shabtai et al. [19] introduce a new framework which extracts execution features and applies Machine Learning techniques to classify them. In a similar way, Arp et al [1] combine different classifiers with execution features creating a new tool, named Drebin. Another similar example is CooperDroid [20] which is focused on reconstructing the behaviour of an Android Application in order to identify Malware. CopperDroid is a dynamic analysis tool focused on analysing system calls, which make the application less vulnerable to alterations. Other example is presented by Isohara et al. [9] which focus on Android markets and its security during the apps validation process. This work shows that some markets are not focused on detecting malicious apps. This compromise several devices and sensitive data. Authors create an audit framework to monitor the application behaviour. They record all the system calls and use signatures of normal system calls. The technique used to detect the Malware behaviour is based on kernels. DroidSIFT [21] proposes a similar semantic-based methodology, based on API Dependency Graphs. Authors use a similarity metric between the API Dependency Graphs of different applications and a threshold to detect anomalies. Also their method aims to detect different malware families.

This work aims to present a detection methodology based on classification that requires non static or dynamic analysis.

3   Modelling the Feature Space

Android shares a lot of features with Java, where the most relevant is its portability. This portability is translated to a Virtual Machine system that interprets bytecode (also known as dex code in Android devices) contained in different files. One of the main differences between Android and Windows malware is the universality of this bytecode, making Android binaries more representatives in a general way than Windows binaries.

3.1   Feature Selection

Each Android app, at the string level, have two representative factors to model: its bytecode and its permission policies. Due to normally Anti-Virus work in binary level we want to perform the comparison in this level including no decompilation or running to the analysis.

The information that we use to model a program is the 1-gram frequency and the permission policies. Using these features we train a classifier in order to measure the ability of the trained classifier to detect malware. This classifier will be compare with AV engines to compare our approach with commercial solutions.

3.2 Classification Process

The classification process is based on discriminating malware and benign-ware using a learning process which aims to identify patterns from different permission policies and instructions used. In this case, we set two obvious classes: malware and benign-ware. This information feeds the classifier in order to learn the main differences among them. The following classifiers have been selected in order to perform this task:

- **Support Vector Machines (SVM)**[3]: This method usually changes the dimension of the search space through different kernel functions, while trying to improve the classification through a hyperplane separation of the data instances in the expanded space [6]. The kernel that has been applied with this function is Radial Basis Function (RBF), this can determine when the separation is linear or not.
- **Inference Trees (IT)**[4]: It divides the data linearly using limits in the attributes and generating a decision tree. The division is chosen using a metric, in this case, the data entropy [15].
- **Random Forest (RF)**[5]: It is a hybrid method that incorporates the advantages of combine different tree classifiers. This methodology trains the several trees and assigns a confidence value to each tree, creating a voting system. This confidence value is used to reach an agreement between the different tree classifiers [7]. It helps to determine when there are sections that are not totally linear.
- **Recursive Partition (RP)**[6]: It is based on inference trees. It continues cutting the tree recursively, to improve the separation. It is helpful to determine when the boundary is not linear.
- **Bagging (B)**[7] This methodology combines different classifiers using a voting system in a similar way to random forest. The chosen classifiers are Recursive Partition classifiers in order to compare with Random Forest, which uses inference trees.
- **Naïve bayes (NB)**[8]: Nave Bayes (NB) is based on Bayes Probability Laws and considers each feature independently of the rest [5]. Each feature contributes to the model information. This helps to understand when the features

---

[3] https://cran.r-project.org/web/packages/e1071/index.html
[4] https://cran.r-project.org/web/packages/party/index.html
[5] https://cran.r-project.org/web/packages/party/index.html
[6] https://cran.r-project.org/web/packages/rpart/index.html
[7] https://cran.r-project.org/web/packages/ipred/index.html
[8] https://cran.r-project.org/web/packages/e1071/index.html

are independent or they need to complement their information with another feature.

The classifier will provide a good discrimination model. Once the model is generated, we test it using fresh data to measure its quality. Due to the classification is a statistical process, we have performed this operation several times, in order to provide a distribution of the accuracy values.

4 Experimental Setup

In order to provide some experimental evidences, we aim to answer the following Research Questions:

- **RQ1: What detection and accuracy can be achieved by the classifiers during the discrimination process?** This research question aims to measure the different performance of the classifiers in order to evaluate whether they can be used as Android malware detectors or not.
- **RQ2: What are the improvements compared with AV Engines?** Here, a comparison with commercial AV Engines is performed in order to evaluate how our methodology improves the commercial solutions.

4.1 Dataset

To validate our detectors, we have extracted malicious and benign Android apks. The malicious apks have been taken from VirusShare [9] and the benign apks from Aptoide[10]. Each apk has been unzipped and its dex files have been transformed to class files. After, they have been concatenated for the 1-grams extraction process. All the apks has been reported using VirusTotal API [11]. VirusTotal provides a diagnosis of the apks using around 56 different AV Engines. All these engines will be used for the final comparison.

The number of features for the 1-gram extraction is 256 and 139 for the permission policies. The permissions have been modelled using a binary representation where each value corresponds to a specific permission extracted from the manifest. This value is 1 or 0 depending whether the permission is used by the application or not.

4.2 Experimental Setup

The algorithms that have been used for classification process have been extracted from different R packages. In order to avoid over-fitting during the training process, we have set the default parameters of each package.

The experiments have been run 100 times, using different seeds to provide the probability distribution of the solutions. Every experiment uses 2/3 of the

---

[9] http://virusshare.com/
[10] http://www.aptoide.com/
[11] http://www.virustotal.com/

data for training and 1/3 for testing. Only testing accuracy is reported in the following results.

The preprocessing process eliminates those data instances with missing information and reduce the number of features whose correlations are higher than 0.85 according to the Pearson correlation.

The evaluation has been performed using the accuracy value, defined by:

$$ACC = \frac{TP + TN}{TP + TN + FP + FN}, \tag{1}$$

where TP represents the True Positives (malware detected), FP represents the False Positives (benign-ware classified as malware), TN represents the True Negatives (benign-ware classified as benign-ware) and FN represents the False Negatives (malware not detected).

For the evaluation of the detection levels using different False Positive ratios, we have used the detection ratio, defined by the True Positives as:

$$DET = \frac{TP}{TP + FN}. \tag{2}$$

These two metrics are ranged between 0 and 1.

5   Experiments

In this section we present the experimental results produced by the classifiers and we compared them with the rest of the AV Engines.

5.1   Classification Results

The evaluation of the different classifiers is shown in Table 1. This table shows the whole statistics for the classifiers. The results show that Naïve Bayes is the worst classifier obtaining a Median accuracy of 0.8312 and 0.8309 of Mean. This result suggests that the search space generated by the permissions and 1-grams is not a linear space, but the classifier obtains competitive results for the discrimination process as we will see in the following section.

SVM obtains better results, close to 0.9 in Mean and Median, this suggests that the discrimination can be perform in a n-dimensional space using a hyperplane, however, Iteration Trees and Recursive Partition (which are tree based classifiers) obtain better results, suggesting that the features can discriminate the data in a competitive way using only a rules-based discrimination.

The multi-learners approaches (Random Forest and Bagging combined with Recursive Partition) show that the different sections of the search space need special learners in order to perform a perfect discrimination. This is specially highlighted by Bagging classifier, which obtains an accuracy of 0.9745 and 0.9742 in Median and Mean, respectively.

In order to measure the quality of the classifiers with different False Positive rate toleration, which is frequent in Malware Analysis, we have performed a study of the ROC curve of the different classifiers, which is shown in Figure 1.

Classifier	Min	Max	Median	Mean	SD
Naïve Bayes	0.8083	0.8560	0.8312	0.8309	0.0079
Support Vector Machine	0.8845	0.9067	0.8959	0.8958	0.0047
Iteration Tress	0.9454	0.9646	0.9553	0.9552	0.0038
Random Forest	0.9430	0.9664	0.9563	0.9565	0.0046
Recursive Partition	0.9244	0.9526	0.9428	0.9420	0.0058
Bagging	**0.9676**	**0.9799**	**0.9745**	**0.9742**	**0.0024**

Table 1. Accuracy results for the 100 executions. This shows the minimum, maximum, median, mean and standard deviation of the different classifiers.

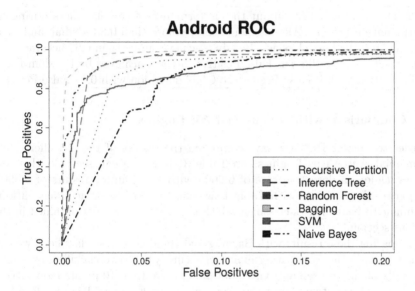

Fig. 1. ROC curve generated by the different classifiers for the Android data.

We can see that only three classifiers are able to obtain detection ratios with 0 False Positives: SVM, Recursive Partition and Random Forest. Besides, the results achieved by Random Forest are really competitive for the 0 False Positive ratio (detection of 0.4065).

The rest of the classifiers start to detect malware when the False Positive ratio is close to 0.01. From this specific value, the best classifier is Bagging, achieving ratios from 0.8953 with 0.01 False Positives to 0.9988 with 0.15 False Positives. This shows that the classifier is able to detect almost all the malware with only 0.15 False Positives. Random Forest is the second best classifier of the sequence, achieving a maximum detection of 0.9977 with 0.15 False Positives. The rest of the classifiers also achieve detection results over 0.92 when the False Positives are close to 0.15.

Classifier	0.0	0.001	0.01	0.05	0.1	0.15
Bagging	0.0	0.0	**0.8953**	**0.9880**	**0.9958**	**0.9988**
Support Vector Machine	0.0216	0.0896	0.6741	0.8262	0.8941	0.9230
Iteration Tress	0.0	0.0	0.1203	*0.9627*	0.9735	0.9771
Random Forest	**0.4065**	**0.4203**	*0.7913*	0.9621	*0.9928*	*0.9970*
Recursive Partition	*0.0324*	*0.0324*	0.0324	0.8671	0.9447	0.9627
Naïve Bayes	0.0	0.0	0.0	0.6963	0.9405	0.9778

Table 2. Median different detection ratios depending on the False Positives tolerated by the classifiers.

Research question 1 asks about the performance of the classifiers during the discrimination process. All classifiers achieve more than 0.83 Median and Mean accuracy and the best classifier, Bagging, achieves more than 0.97 accuracy. The best detection with different False Positive ratios is achieved by Random Forest: 0.40 detection with 0 False Positive to 0.997 detection with 0.15 False Positives.

5.2 Comparison with commercial AV Engines

In order to answer RQ2 we have compared the results of the classifiers with commercial Anti-Virus Engines. As Table 3 shows we can see that the best engines achieve a detection rate of 0.6606 with an accuracy results of 0.8303. Comparing these results with Table 1, we can see that these accuracy rations are similar to Naïve Bayes classifier, which is the worst classifier obtained in the previous section.

Comparing these results with Bagging and Random Forest classifiers, we can see that, even in the worst case (minimum accuracy) the two classifiers overcome the results of the top ten anti-virus, obtaining more than 10 points (considering the values as percentages) over the best Anti-Virus Engine: 0.9430 for Random Forest and 0.9676 for Bagging, compared with 0.8303 for Cyren, the best AV engine.

Engine	Detection	Accuracy
Cyren	0.6606	0.8303
Ikarus	0.6544	0.8272
VIPRE	0.6506	0.8253
McAfee	0.649	0.8245
AVG	0.6472	0.8236
AVware	0.639	0.8195
ESET.NOD32	0.6362	0.8181
CAT.QuickHeal	0.6358	0.8179
AegisLab	0.6348	0.8174
NANO.Antivirus	0.623	0.8115

Table 3. Top ten accuracy and detection results for the 56 different commercial engines applied by VirusTotal.

Research question 2 asks about the improvements of the classifier compared with commercial Anti-Virus engines. The results have shown that the worst classifier, Naïve Bayes, obtains similar results than the best AV Engine and the best classifier, Bagging, obtains more than 10 points (0.97 compared with 0.83) over the best commercial engine.

6 Conclusions and Future Work

This work presents a straightforward methodology to extract features from Android apks and uses these features to discriminate malware. This process extracts the 1-gram structure of the dex files and generates a frequency representation which is related to the low level opcodes operations performed by the software. Joining this information with the app permission policy, the systems is able to discriminate malware and benign-ware with no running or disassembly process, which are frequent and expensive methodologies in static and dynamic program analysis, respectively.

The results show that this approach overcomes the discrimination of commercial Anti-Virus Engines in all cases and significantly in the best case. Besides, the detection levels achieved by the best classifiers are closed to the total detection when the False Positive ratio is only 0.15.

The future work will be focused on incrementing the precision of this methodology trying to include other features such as third party API calls or meta-information of the applications.

7 Acknowledgements

This work has been supported by the next research projects: Spanish Ministry of Economy and Competitivity and European Regional Development Fund FEDER (TIN2014-56494-C4-4-P), CIBERDINE S2013/ICE-3095, SeMaMatch EP/K032623/1 and Airbus Defence & Space (FUAM-076914 and FUAM-076915).

References

1. Daniel Arp, Michael Spreitzenbarth, Malte Hübner, Hugo Gascon, Konrad Rieck, and CERT Siemens. Drebin: Effective and explainable detection of android malware in your pocket. In *Proceedings of the Annual Symposium on Network and Distributed System Security (NDSS)*, 2014.
2. Gema Bello-Orgaz and David Camacho. Evolutionary clustering algorithm for community detection using graph-based information. In *Evolutionary Computation (CEC), 2014 IEEE Congress on*, pages 930–937. IEEE, 2014.
3. Gema Bello-Orgaz, Jason J Jung, and David Camacho. Social big data: Recent achievements and new challenges. *Information Fusion*, 28:45–59, 2016.
4. Mihai Christodorescu, Somesh Jha, Sanjit Seshia, Dawn Song, Randal E Bryant, et al. Semantics-aware malware detection. In *Security and Privacy, 2005 IEEE Symposium on*, pages 32–46. IEEE, 2005.

5. Pedro Domingos and Michael Pazzani. On the optimality of the simple bayesian classifier under zero-one loss. *Machine learning*, 29(2-3):103–130, 1997.

6. Marti A. Hearst, Susan T Dumais, Edgar Osman, John Platt, and Bernhard Scholkopf. Support vector machines. *Intelligent Systems and their Applications, IEEE*, 13(4):18–28, 1998.

7. Tin Kam Ho. The random subspace method for constructing decision forests. *Pattern Analysis and Machine Intelligence, IEEE Transactions on*, 20(8):832–844, 1998.

8. Nwokedi Idika and Aditya P Mathur. A survey of malware detection techniques. *Purdue University*, 48, 2007.

9. Takamasa Isohara, Keisuke Takemori, and Ayumu Kubota. Kernel-based behavior analysis for android malware detection. In *Computational Intelligence and Security (CIS), 2011 Seventh International Conference on*, pages 1011–1015. IEEE, 2011.

10. Daniel T Larose. *Discovering knowledge in data: an introduction to data mining*. John Wiley & Sons, 2014.

11. Hector D Menendez, David F Barrero, and David Camacho. A genetic graph-based approach for partitional clustering. *International journal of neural systems*, 24(03):1430008, 2014.

12. Héctor David Menéndez and David Camacho. Mogcla: A multi-objective genetic clustering algorithm for large data analysis. In *Proceedings of the Companion Publication of the 2015 on Genetic and Evolutionary Computation Conference*, pages 1437–1438. ACM, 2015.

13. Andreas Moser, Christopher Kruegel, and Engin Kirda. Limits of static analysis for malware detection. In *Computer security applications conference, 2007. ACSAC 2007. Twenty-third annual*, pages 421–430. IEEE, 2007.

14. Mila Dalla Preda, Mihai Christodorescu, Somesh Jha, and Saumya Debray. A semantics-based approach to malware detection. *ACM SIGPLAN Notices*, 42(1):377–388, 2007.

15. J Ross Quinlan and Ronald L Rivest. Inferring decision trees using the minimum description lenght principle. *Information and computation*, 80(3):227–248, 1989.

16. Víctor Rodríguez-Fernández, Héctor D Menéndez, and David Camacho. Automatic profile generation for uav operators using a simulation-based training environment. *Progress in Artificial Intelligence*, 5(1):37–46, 2016.

17. Victor Rodriguez-Fernandez, Cristian Ramirez-Atencia, and David Camacho. A multi-uav mission planning videogame-based framework for player analysis. In *Evolutionary Computation (CEC), 2015 IEEE Congress on*, pages 1490–1497. IEEE, 2015.

18. Igor Santos, Felix Brezo, Javier Nieves, Yoseba K Penya, Borja Sanz, Carlos Laorden, and Pablo G Bringas. Idea: Opcode-sequence-based malware detection. In *Engineering Secure Software and Systems*, pages 35–43. Springer, 2010.

19. Asaf Shabtai, Uri Kanonov, Yuval Elovici, Chanan Glezer, and Yael Weiss. andromaly: a behavioral malware detection framework for android devices. *Journal of Intelligent Information Systems*, 38(1):161–190, 2012.

20. Kimberly Tam, Salahuddin J Khan, Aristide Fattori, and Lorenzo Cavallaro. Copperdroid: Automatic reconstruction of android malware behaviors. In *Proc. of the Symposium on Network and Distributed System Security (NDSS)*, 2015.

21. Mu Zhang, Yue Duan, Heng Yin, and Zhiruo Zhao. Semantics-aware android malware classification using weighted contextual api dependency graphs. In *Proceedings of the 2014 ACM SIGSAC Conference on Computer and Communications Security*, pages 1105–1116. ACM, 2014.

Part IV

Space-Based Coordination

Optimal Configuration Model of a Fleet of Unmanned Vehicles for Interoperable Missions

Gabriella Gigante[1], Domenico Pascarella[1], Salvatore Luongo[1], Carlo Di Benedetto[1], Angela Vozella[1], and Giuseppe Persechino[2]

[1] Integrated Software - Verification and Validation Laboratory, CIRA (Italian Aerospace Research Centre), Capua, Italy,
g.gigante@cira.it, d.pascarella@cira.it, s.luongo@cira.it,
c.dibenedetto@cira.it, a.vozella@cira.it
[2] Land, Environment and Cultural Assets Department, CIRA (Italian Aerospace Research Centre), Capua, Italy,
g.persechino@cira.it

Abstract. It is largely recognized that many missions may be easily performed by unmanned vehicles both in military and in civil domain. Literature shows a large inventory of their applications with operational and logistical challenges. Comparing different types of missions, a multi-vehicle approach is able to guarantee better performances and minimum costs, as long as they are coordinated. Thus, the problem to guarantee the better platform configuration to perform the mission becomes architecting the best fleet. This paper proposes an approach to identify the best fleet to perform an envelope of missions, by transforming the architecting activity in an optimization problem. A classification of unmanned vehicles missions and the formal definition of the problem are proposed.

Keywords: Unmanned vehicles, Fleet composition, Surveillance

1 Introduction

Unmanned Vehicles (UVs) have been used in the past years over a wide range of military missions. Nevertheless, they currently represent an alternative to manned vehicles for a relevant number of civilian applications. Overcoming the initial technology issues, the literature shows a large inventory of applicative research. A single powerful UV equipped with a large array of different sensors is limited to a single point of view. In the last years, the multi-UV approach seems to be more suitable approach for many applications since it can guarantee:

- Multiple simultaneous interventions – A multi-UV team, or Multi Robot System (MRS), can simultaneously collect data from multiple locations;
- Greater efficiency – A multi-UV team can split up in order to efficiently cover a large area, optimizing available resources.
- Complementarity – A multi-UV team can perform different tasks with growing accuracy. For example, a stratospheric platform can detect a possible target, a fixed wing vehicle can perform the identification of the target, a ground vehicle can perform the reconnaissance.

© Springer International Publishing AG 2017   111
C. Badica et al. (eds.), *Intelligent Distributed Computing X*,
Studies in Computational Intelligence 678, DOI 10.1007/978-3-319-48829-5_11

- Reliability – A multi-UV team assures fault tolerant missions by providing redundancy and capability of reconfiguration in the case of a failure of a vehicle.
- Safety – For a permit to fly, a fleet of mini-UAVs is safer than a single great and heavy UAV.
- Cost Efficiency – A single vehicle to execute some tasks could be an expensive solution when comparing to several low cost vehicles.

In futuristic scenarios, it is conceivable to have available interoperable UVs and payloads in order to identify the best team to perform a given mission. The result of this reasoning has led to the current work. It proposes a possible approach at the basis of a framework able to identify the best team of UVs with the most suitable sensors to achieve mission objectives, under some basic hypotheses. The architecture of multiple UVs system is designed to minimize the needed costs, while respecting the mission performance constraints. The definition of the system architecture is turned to an optimization problem with the definition of the formal model, of the constraints, and of the objective function. The problem has three basic classes of elements: the mission, the UVs, and the possible payloads. To define a formal model of the problem, it is necessary to model all the previous elements. A classification of missions with particular focus on those requiring interoperability in terms also of the related commonalities enables the identification of possible paradigms. Paradigms identify the relative characterizing aspects that in turn are the basic concepts to model the mission. Constraints express the suitability of the vehicle against the payload, of the payload against the target, and of the vehicle against the target. Conclusion by proposing possible solutions to the problem are described.

2   Related Work

The problem related to the optimization of the fleet is quite new. Literature on multi-UV approach addresses at least three main research directions: interoperability, mission planning and applicative case studies.

Interoperability is defined in the glossary of the Defense Acquisition University as: "*The ability of systems, units, or forces to provide data, information, materiel, and services to and accept the same from other systems, units, or forces and to use the data, information, materiel, and services so exchanged to enable them to operate effectively together*" [1].

There exist different frameworks describing interoperability dimensions (the most technical ones and those related to organizations, business, etc.). The technical level of interoperability could be mainly related to the way UVs perform the mission together and the way each single UV is defined to enable the working in team. Dimensions are cooperation, coordination and collaboration, the knowledge level, communication, team composition and size and system architecture. Besides, the organizational level requiring different entities to work together could be seen as the way the decision system is organized. Literature provides different definition of cooperation. It can be defined as a joint behavior to operate together that is directed toward some goal in which there is a common interest. Collaboration is defined as *"an integrated behavior directed toward the same goal in which there is a common interest"*. The term "joint" means that all vehicles have the same skill and share their resources to achieve the mission goal minimizing time or costs. The term "integrated" means that platforms have complementary skills and that their integration enables the achievement of a goal that individually could not reach.

The team composition and size are also other important issues for multi-UV approach [6].Mission planning means the activity elaborated by a central or by a distributed organization allowing the definition and the allocation of tasks. To perform mission planning, it is primarily necessary divide the mission in tasks and then allocate such tasks to the proper vehicles. Taxonomies in both case come into help. From one hand they can allow a uniform identification of tasks across the large inventory of missions. A detailed study is reported in next section. From the other hand they guide the study of the problem of the tasks allocation.

A recent work focusing on the analysis of task allocation in MRS is presented in [3]. This work proposes a formal framework for the study of this problem. The authors consider three main dimensions: Single-Task (ST) vs Multi-Task (MT) robots, based on whether the robots involved in the task assignment process can execute more than one task at a time; Single-Robot (SR) vs Multi-Robot (MR) tasks, considering if the tasks to be performed involve one or multiple robots; Instantaneous Assignment (IA) vs Time-extended Assignment (TA), distinguishing whether the information concerning the robots tasks and environment permit only an instantaneous assignment or a more sophisticated planning approach.

The distributed allocation of tasks refers to the study of the problem of assigning the task to a group of agents in a distributed manner. It contains the problem of coverage control, which formally requires the full coverage of an area. Much of the research on the subject is reported in [7]. The task allocation also includes the distributed scheduling problem.

3   Background

To define a formal model of the problem, it is necessary to model the mission, the vehicles and the payloads.

To model a mission, the first step is an initial classifications of applications. In particular, three main types of applications are distinguished: government, which are managed by government agencies; civilian, which are the civilian that are not properly handled by government agencies; military. For each of these classes, it is possible to define sub-classes. For example, the government applications can be divided in security and search and rescue.

Some mission characteristics can be assembled in mission paradigms, whose attributes are such that aggregate multiple similar applications. In this paper, we will consider observation missions, which require the location of a target entity in a region of interest. The main identified paradigms within this category are: persistent surveillance, which is the set of missions in which n platforms must reach m geographical coordinates (with $n < m$) and monitor them; coverage, that is the set of missions where one or more platforms efficiently move in a region in order to have a complete coverage of the area; exploration, which is the set of missions where one or more platforms efficiently move in a region in search of one or more fixed targets; pushing invaders, which is the set of missions where one or more platforms efficiently move in a region in search of one or more fixed targets; target tracking, which is the set of missions where one or more platforms aim to pursue one or more stationary or moving targets in a region.

Many missions require the use of more paradigms in order to cover all the mission objectives. Characterizing features for the specific paradigm are the set of tasks, the mission duration, the region of interest, the required pixel density, bandwidth and accessibility for the task observation and the required task refresh time. Each of these paradigms may involve one or more agents, depending on the required performance. The inspection accuracy depends on the characteristics of the pairs vehicles – boarded sensors.

The platforms are unmanned vehicles, which are piloted by remote or otherwise operating in automatic or autonomous mode. Based on the operating environment (air, space, ground and water), the following types of drones can be distinguished: UAV (Unmanned Aerial Vehicle), operating in the air; satellites, operating in space; UGV (Unmanned Ground Vehicle), operating on the ground; UMV (Unmanned Marine Vehicle), operating in water. According to the proposed model, the main attributes are the altitude, the maximum and minimum speed, the endurance, the type, the operational and consumption costs and the payload weight capacity.

Regarding the payloads, their distinctive features for an observation mission are the resolution, the bandwidth, the field of view and the weight. About the payload field of view, we consider the case of a stratospheric platform as an example. Here, the acquisition may be performed both in "across track scanner" mode (i.e., along the orthogonal direction with respect to the ground track) and in "along track scanner mode" (i.e., along the ground track), as shown in Fig. 1.

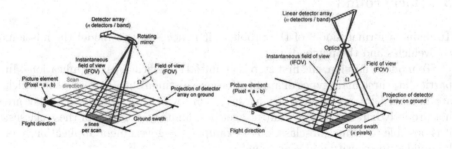

Fig. 1. Along (left) and across (right) track scanner.

4   Formal Statement of the Problem

The following sections highlight the formal model of the proposed problem.

4.1   Mission Model

A mission M may be defined as the following tuple

$$M = \langle \mathbb{T}, P_a, T_m \rangle \ . \tag{1}$$

P_a is the identifier of the mission paradigm (e.g., persistent surveillance, exploration, etc.). In the following, we will consider the specific model for a persistent surveillance mission. $T_m \in \mathbb{R}^+$ is the mission time. $\mathbb{T}$ is the set of tasks. If $N_t \in \mathbb{N}^*$ is the number of tasks, $\mathbb{T}$ is defined as

$$\mathbb{T} = \{\mathbf{t}_1, \dots, \mathbf{t}_{N_t}\} \ . \tag{2}$$

Each task $\mathbf{t}_i$ is a tuple that has the following expression

$$\mathbf{t}_i = \left\langle \rho_t^{(i)}, \mathbf{s}_t^{(i)}, \mathbb{B}_t^{(i)}, T_r^{(i)}, T_o^{(i)}, A_t^{(i)}, M_o^{(i)}, ROI_t^{(i)} \right\rangle, \quad \forall i \in (1, \dots, N_t) \ . \tag{3}$$

The term $\rho_t^{(i)} \in \mathbb{R}^+$ is the required pixel density. The term $\mathbf{s}_t^{(i)} \in \mathbb{R}^2$ is the two-dimensional central position of the task. The term $T_r^{(i)} \in \mathbb{R}^+$ is the maximum allowable refresh time for the task, whereas the term $T_o^{(i)} \in \mathbb{R}^+$ represents the required observation time for the task. The term $\mathbb{B}_t^{(i)}$ is the required bandwidth for the task, which is defined as

$$\mathbb{B}_t^{(i)} = \left\{ \left[B_{t_{\inf}}^{(i)}, B_{t_{\sup}}^{(i)} \right] \mid B_{t_{\inf}}^{(i)} \in \mathbb{R}^+, \ B_{t_{\sup}}^{(i)} \in \mathbb{R}^+, \ B_{t_{\inf}}^{(i)} < B_{t_{\sup}}^{(i)} \right\},$$
$$\forall i \in (1, \dots, N_t) \ . \tag{4}$$

The term $A_t^{(i)}$ is the possible accessibility for the task, which is defined as

$$A_t^{(i)} \in \{water, ground, air, (ground, air), (water, air)\}, \ \forall i \in (1, \dots, N_t) \ . \tag{5}$$

The term $M_o^{(i)}$ is the required observation manoeuvre for the task, which has the following expression

$$M_o^{(i)} \in \{loitering, \ hovering\}, \quad \forall i \in (1, \dots, N_t) \ . \tag{6}$$

The term $\mathbb{ROI}_t^{(i)}$ is the region of interest of the task, which is

$$\mathbb{ROI}_t^{(i)} = \left\{ \left[ROI_{t_{\inf}}^{(i)}, ROI_{t_{\sup}}^{(i)} \right] \mid ROI_{t_{\inf}}^{(i)} \in \mathbb{R}^+, \ ROI_{t_{\sup}}^{(i)} \in \mathbb{R}^+, \right.$$
$$\left. ROI_{t_{\inf}}^{(i)} < ROI_{t_{\sup}}^{(i)} \right\}, \quad \forall i \in (1, \dots, N_t) \ . \tag{7}$$

4.2   Vehicles and Payloads Model

Let $N_v \in \mathbb{N}^*$ be the number of vehicles. The set of vehicles $\mathbb{V}$ in the fleet is defined as

$$\mathbb{V} = \{\mathbf{v}_1, \dots, \mathbf{v}_{N_v}\} \ . \tag{8}$$

Each vehicle $\mathbf{v}_i$ is defined as the following tuple

$$\mathbf{v}_i = \left\langle \mathbf{s}_v^{(i)}, h^{(i)}, E^{(i)}, v_{\min}^{(i)}, v_{\max}^{(i)}, A_v^{(i)}, W_v^{(i)}, C_o^{(i)}, C_c^{(i)} \right\rangle, \quad \forall i \in (1, \dots, N_v) \ . \tag{9}$$

The instantaneous position $\mathbf{s}_v^{(i)} \in \mathbb{R}^2$ of the vehicle is its projected position along the two-dimensional plane of the region of interest. The term $h^{(i)} \in \mathbb{R}^+$ is the reference altitude of the vehicle. The term $E^{(i)} \in \mathbb{R}^+$ is the endurance of the vehicle. The term $v_{\min}^{(i)} \in \mathbb{R}_0^+$ is the minimum speed of the vehicle: if it is null, the vehicle is allowed to hover. The term $v_{\max}^{(i)} \in \mathbb{R}^+$ (with $v_{\min}^{(i)} \le v_{\max}^{(i)}$) is the maximum speed of the vehicle. The term $A_v^{(i)}$ represents the type of UV, i.e.

$$A_v^{(i)} \in \{water, \ ground, \ air\}, \quad \forall i \in (1, \dots, N_v) \ . \tag{10}$$

The term $W_v^{(i)} \in \mathbb{R}^+$ quantifies the allowable payload weight of the vehicle. The term $C_o^{(i)} \in \mathbb{R}_0^+$ is the fixed operational cost (i.e., the invariable cost for the vehicle usage in the mission). The term $C_c^{(i)} \in \mathbb{R}_0^+$ is the vehicle consumption cost, which is a constant related to the consumption model of the vehicle.

Let $N_p \in \mathbb{N}^*$ be the number of payloads. The set of payloads $\mathbb{P}$ is defined as

$$\mathbb{P} = \left\{ \mathbf{p}_1, \dots, \mathbf{p}_{N_p} \right\} \ . \tag{11}$$

Each payload is defined as the following tuple

$$\mathbf{p}_i = \left\langle R_p^{(i)}, FoV^{(i)}, \mathbb{B}_p^{(i)}, W_p^{(i)} \right\rangle, \quad \forall i \in (1, \dots, N_p) \ . \tag{12}$$

The term $R_p^{(i)} \in \mathbb{R}^+$ is the resolution of the payload. The term $FoV^{(i)} \in \mathbb{R}^+$ is the field of view of the payload. The term $W_p^{(i)} \in \mathbb{R}^+$ is the payload weight. The term $\mathbb{B}_p^{(i)}$ is the provided bandwidth of the payload, i.e.

$$\mathbb{B}_p^{(i)} = \left\{ \left[B_{p_{\inf}}^{(i)}, B_{p_{\sup}}^{(i)} \right] \mid B_{p_{\inf}}^{(i)} \in \mathbb{R}^+, \ B_{p_{\sup}}^{(i)} \in \mathbb{R}^+, \ B_{p_{\inf}}^{(i)} < B_{p_{\sup}}^{(i)} \right\}, \tag{13}$$
$$\forall i \in (1, \dots, N_p) \ .$$

4.3   Problem Model

The stated problem: is related to an observation mission; considers multi-task and multi-robot paradigms; takes into account both cooperation and collaboration. Moreover, the following hypotheses are assumed: the tasks are fully known and stationary; the fleet is all available at the position $\mathbf{s}_0$; the recharging/refueling of the vehicles is not considered.

Let $\mathbb{F}$ be the fleet configuration, i.e., the set of vehicle configurations for a given mission M. Hence, $\mathbb{F}$ has the following expression

$$\mathbb{F} = \{\mathbf{f}_1, \ldots, \mathbf{f}_{N_v}\} \tag{14}$$

The term $\mathbf{f}_i$ is the configuration of the i-th vehicle.

$$\mathbf{f}_i = \left\langle \mathbb{P}^{(i)}, \mathcal{S}^{(i)} \right\rangle, \quad \forall i \in (1, \ldots, N_v) \tag{15}$$

$\mathbb{P}^{(i)} \subseteq \mathbb{P}$ is the set of payloads that are assigned to the vehicle $\mathbf{v}_i$ by means of the configuration $\mathbf{f}_i$. $\mathcal{S}^{(i)} \in 2^{\mathbf{T}^\infty}$ is the plan (i.e., the sequence of tasks) that is assigned to the vehicle $\mathbf{v}_i$ by means of the configuration $\mathbf{f}_i$.

The proposed problem of the optimal assignment of a fleet configuration for the mission M is defined as

$$\mathbb{F}_{opt} = \left\langle \mathbb{P}^{(i)_{opt}}, \mathcal{S}^{(i)_{opt}} \right\rangle_{|i=1,\ldots,N_v} = \underset{\mathbb{F}=\{\mathbf{f}_i\}_{|i=1,\ldots,N_v}}{\text{argmin}} \sum_{i=1}^{N_v} C_v^{(i)}(\mathbf{f}_i) . \tag{16}$$

Note that the configuration problem in (16) may be seen as the set of two sub-problems: the composition problem, i.e., the selection of the vehicles and the respective boarded payloads for the accomplishment of M; the routing problem, i.e., the assignment of the sequence of tasks to each vehicle.

The term $C_v^{(i)}(\mathbf{f}_i)$ is the configuration cost of the i-th vehicle, which is defined as

$$C_v^{(i)}(\mathbf{f}_i) = \begin{cases} 0, & \text{if } \mathcal{S}^{(i)} = \emptyset \\ C_o^{(i)} + C_{plan}\left(C_c^{(i)}, \mathbb{P}^{(i)}, \mathcal{S}^{(i)}\right), & \text{if } \mathcal{S}^{(i)} \neq \emptyset \end{cases}, \; i \in (1, \ldots, N_v). \tag{17}$$

Thus, the configuration cost (if the vehicle is employed in the mission) is the sum of two terms. The first is a linear term, which represents the fixed (operational) vehicle cost. The second is a non-linear term, which is generically a function of the consumption cost of the vehicle, of the weight and of the plan (i.e., it is intuitively the sum of the en route costs, which are related to the routing graph of the tasks).

Problem (16) shall be subject to the following constraints:

- Starting position of the vehicles – The employed vehicles shall start from a given position $\mathbf{s}_0 \in \mathbb{R}^2$, i.e.

$$\mathbf{s}_v^{(i)}(0) = \mathbf{s}_0, \quad \mathbf{s}_0 \in \mathbb{R}^2, \; \forall i \in (1, \ldots, N_v): \; \mathcal{S}^{(i)} \neq \emptyset . \tag{18}$$

- Final position of the vehicles – The employed vehicles shall finish in the starting position $\mathbf{s}_0$, i.e.

$$\mathbf{s}_v^{(i)}(t) = \mathbf{s}_0, \quad \mathbf{s}_0 \in \mathbb{R}^2, \; t \geq T_m, \; \forall i \in (1, \ldots, N_v): \; \mathcal{S}^{(i)} \neq \emptyset . \tag{19}$$

- Compatibility vehicle-mission (endurance) – The endurance of each employed vehicle shall be compatible with the required mission time, i.e.

$$T_m \leq E_{sf} \cdot E^{(i)}, \quad \forall i \in (1, \dots, N_v): \; \mathcal{S}^{(i)} \neq \emptyset \; . \tag{20}$$

wherein $E_{sf} \in (0, 1]$ is the endurance safety margin.
- Compatibility vehicle-task (accessibility) – The accessibility of each employed vehicle shall be compatible with the required accessibility, i.e.

$$\mathbf{t}_j \in \mathcal{S}^{(i)} \implies A_v^{(i)} \subseteq A_t^{(j)}, \quad \forall j (1, \dots, N_t), \; i \in (1, \dots, N_v) \; . \tag{21}$$

- Compatibility vehicle-task (observation manoeuvre) – The required observation manoeuvre shall be compatible with the dynamic features of the vehicle, i.e.

$$\mathbf{t}_j \in \mathcal{S}^{(i)} \text{ and } M_o^{(j)} = hovering \implies v_{\min}^{(i)} = 0,$$
$$\forall j (1, \dots, N_t), \; i \in (1, \dots, N_v) \; . \tag{22}$$

- Payload uniqueness – Each employed payload shall be assigned to a single vehicle, i.e.

$$\mathbf{p}_k \in \mathbb{P}^{(i)} \implies \mathbf{p}_k \notin \mathbb{P}^{(j)}, \; \forall k \in (1, \dots, N_p), \; i, j \in (1, \dots, N_v), \; i \neq j \; . \tag{23}$$

- Compatibility vehicle-payload (weight) – The boarded payloads shall respect the weight constraint of the assigned vehicle, i.e.

$$\sum_{\mathbf{p}_j \in \mathbb{P}^{(i)}} W_p^{(j)} \leq W_{sf} \cdot W_v^{(i)}, \quad \forall i \in (1, \dots, N_v): \; \mathcal{S}^{(i)} \neq \emptyset \; . \tag{24}$$

wherein $W_{sf} \in (0, 1]$ is the weight safety margin.
- Conflicts avoidance – A task shall be assigned to one UV at an instant, i.e.

$$\mathbf{t}_k \in \mathcal{S}^{(i)}(t) \implies \mathbf{t}_k \notin \mathcal{S}^{(j)}(t), \; \forall i, j \in (1, \dots, N_v), \; \forall k (1, \dots, N_t), \; \forall t. \tag{25}$$

- Compatibility payload-task (bandwidth) – The bandwidht of a task shall be included in the bandwidth of at least one boarded payload of the vehicle that is assigned to the task, i.e.

$$\mathbf{t}_j \in \mathcal{S}^{(i)} \implies \exists \, \mathbf{p}_k \in \mathbb{P}^{(i)}: B_t^{(j)} \subseteq B_p^{(k)}, \forall j (1, \dots, N_t), \; i \in (1, \dots, N_v). \tag{26}$$

- Compatibility payload-task (pixel density, resolution, FoV) – If $\mathrm{ROI}_{\mathbf{p}_k}^{(j)} = g\left(h^{(i)}, FoV^{(k)}\right)$ is the ground path density of the payload $\mathbf{p}_k$ for the j-th task and if $\rho_{\mathbf{p}_k}^{(j)} = R_p^{(k)}/\mathrm{ROI}_p^{(j)}$ is the pixel density of the payload $\mathbf{p}_k$ for the j-th task, the following constraint shall hold

$$\mathbf{t}_j \in \mathcal{S}^{(i)} \Rightarrow \exists \mathbf{p}_k \in \mathbb{P}^{(i)}: \rho_p^{(i)} \geq \rho_t^{(i)}, \mathrm{ROI}_{\mathbf{p}_k}^{(j)} \subseteq \mathrm{ROI}_t^{(i)}, \forall i \in (1, \dots, N_v). \tag{27}$$

- Tasks refresh times – Let $\mathcal{S} = \bigcup_{i=1}^{N_v} \mathcal{S}^{(i)}$ the joint plan of all the vehicles in a given configuration and let $T_r^{(j, \mathcal{S})}$ be the maximum refresh time of the task $\mathbf{t}_j$ by means of the joint plan $\mathcal{S}$. The following relation shall hold

$$T_r^{(j, \mathcal{S})} \leq T_r^{(i)}, \quad \forall j (1, \dots, N_t) \; . \tag{28}$$

4.4 Problem Analysis

The stated problem in Sect. 4.3 may be seen as a variant of well-known optimization problems, such as the vehicle routing problems (VRPs) [8]. Basically, the VRP deals with the optimal assignment of a set of transportation orders (i.e., route segments) to a fleet of vehicles and the sequencing of stops for each vehicle in order to serve a given set of customers with known delivery demands. The objective usually reflects the minimization of total transportation costs. For example, reference [5] provides a survey of the mathematical models for the optimization problems of logistics infrastructure.

Anyway, this work concerns an heterogeneous fleet of unmanned vehicles, contrary to the typical assumption of homogeneous fleet of VRP theory. Moreover, it tackles a combined fleet composition and routing problem. Indeed, the proposed fleet configuration problem of equation (16) includes two tasks: the fleet composition, i.e., the choice of the optimal fleet; the fleet routing, i.e., the optimal policy for the tasks allocation. In other words, the fleet composition works on a strategic horizon, whereas the fleet routing regards a tactical horizon. By ignoring routing aspects, fleet composition decisions may be based on a too simplified view on tasks demand.

A first extension of the VRP is the heterogeneous fixed fleet vehicle routing problem (HFFVRP), wherein the fleet size is fixed or bounded by a maximum number, but the vehicles can be of different size and have different fixed and variable costs unlike the classical VRP [4]. Here, the aim is not to construct an optimal fleet, but only to use the different vehicles in a best possible way.

On the contrary, the fleet size and mix vehicle routing problem (FSMVRP) is an extension of the VRP to a heterogeneous fleet and an extension of the objective to include vehicle acquisition and routing costs [4]. It differs from VRP by including the composition of the fleet. A number of vehicle types with different capacities and acquisition costs are given. The objective is to find a fleet composition and a corresponding routing plan that minimizes the sum of the fixed costs for managing the vehicles in the fleet and variable routing costs. Another reference problem for our work is the FSMVRP with time windows (FSMVRPTW). This is a natural extension of the FSMVRP, which introduces time windows for each customer defining an interval wherein customer service has to start [4]. Some heuristics to solve the FSMVRPTW are reported in [4, 2].

The addressed problem also deals with persistent surveillance planning, wherein the observation rates of the tasks shall be respected. This may be handled as a further generalization of the time windows in vehicle routing. Indeed, the time windows shall be somehow periodic. The persistent surveillance routing with prefixed limits on the refresh times has been analyzed by means of the problem for persistent visitation under revisit constraints. The authors in [9] discuss the problem of finding paths that meet these revisit rate. They also present periodicity properties of solutions to the path planning problem and some heuristics.

Note that the VRP belongs to the class of NP-hard optimization problems [4], so, all the derived problems are NP-hard by restriction. Furthermore, the persistent visitation problem is proven to be NP-complete [9].

To the best of our knowledge, this is the first work that joins the fleet size and mix vehicle routing problem with the persistent visitation problem under revisit constraints. A possible solution could consist in an initial procedure to create the optimal fleet mix and then to apply this composition to solve the persistent visitation problem (e.g., by infinitely repeating a periodic cycle).

5 Conclusion and Future Work

This work proposes an architecturing approach to configure an optimal multi-UV fleet for the accomplishment of a mission. The configuration of the fleet is the selection of the vehicles and of the most suitable sensors to achieve mission objectives and as the assignment of the routing plan for each vehicle. A formal statement of the problem is provided with the models of the related actors (mission, UVs and payloads) and the required constraints. A classification of interoperable missions and an analysis of platforms are also produced in order to highlight the aspects that are needed to model the mission. Future work will be about the actual implementation of a solution strategy, which will possibly join some well-known heuristics for the FSMVRP and for the persistent visitation problem under revisit constraints.

References

1. DAU: Glossary of defense acquisition acronyms & terms. Tech. rep., Defense Acquisition University (2011)
2. Dell'Amico, M., Monaci, M., Pagani, C., Vigo, D.: Heuristic Approaches for the Fleet Size and Mix Vehicle Routing Problem with Time Windows. Transportation Science 41(4), 516–526 (11 2007)
3. Gerkey, B., Mataric, J.M.: Multi-robot task allocation: Analyzing the complexity and optimality of key architectures. In: Proceedings of the IEEE International Conference on Robotics and Automation (ICRA 2003). IEEE (9 2003)
4. Hoffa, A., Anderssonb, H., Christiansed, M., Haslec, G., Løkketangena, A.: Industrial aspects and literature survey: fleet composition and routing. Computers & Operations Research 37(12), 2041–2061 (12 2010)
5. Kazakov, A.L., Lempert, A.A.: On Mathematical Models for Optimization Problem of Logistics Infrastructure. International Journal of Artificial Intelligence 13(1), 200–210 (2015)
6. Konolige, K.: Centibots: Large scale robot teams. In: Proocedings of the 2003 NRL Workshop on Multi-Robot Systems (2003)
7. Kwok, A., Martinez, S.: Unicycle coverage control via hybrid modeling. IEEE Transactions on Automatic Control 55(2), 528–532 (2010)
8. Laporte, G.: The vehicle routing problem: an overview of exact and approximate algorithms. European Journal of Operational Research 59(3), 345–358 (6 1992)
9. Las Fargeas, J., Hyun, B., Kabamba, P., Girard, A.: Persistent Visitation under Revisit Constraints. In: Proceedings of 2013 International Conference on Unmanned Aircraft Systems (ICUAS). pp. 952–957. IEEE (5 2013)

Spatial Tuples:
Augmenting Physical Reality with Tuple Spaces

Alessandro Ricci, Mirko Viroli, Andrea Omicini,
Stefano Mariani, Angelo Croatti, and Danilo Pianini

Università di Bologna, Italy
{a.ricci,mirko.viroli,andrea.omicini,s.mariani,a.croatti,danilo.pianini}@unibo.it

Abstract. We introduce Spatial Tuples, an extension of the basic tuple-based model for distributed multi-agent system coordination where *(i)* tuples are conceptually placed in the physical world and possibly move, *(ii)* the behaviour of coordination primitives may depend on the spatial properties of the coordinating agents, and *(iii)* the tuple space can be conceived as a virtual layer augmenting physical reality. Motivated by the needs of mobile augmented-reality applications, Spatial Tuples explicitly aims at supporting space-aware and space-based coordination in agent-based pervasive computing scenarios.

1 Introduction

The widespread diffusion of mobile and wearable technologies, along with the pervasive availability of the Internet, makes it possible to conceive a wide range of distributed, location- and context-aware applications where the actual users' position in the physical space is essential for the system [3, 9, 19, 18].

Coordination plays a key role here, in particular within collaborative situated applications, where multiple users act and interact inside the same physical environment. There, the notion of space can be exploited to design coordination strategies and patterns, based on the positions of users and other situated entities of the system. Two foremost examples are *situated communication* – i.e., messages perceived only by users located in some specific place – and *spatial synchronisation*—i.e., ordering agents' actions based on their physical position.

Accordingly, different kinds of coordination models and technologies were proposed in literature, mainly in the context of mobile computing and *spatial computing* [10, 14, 15]. Most of these approaches were inspired by the LINDA coordination language [7], where distributed agents interact and coordinate by exchanging messages through shared information spaces (called *tuple spaces*), exploiting *generative communication*. These *tuple-based* models provide coordination mechanisms supporting space, time, and reference *uncoupling*, promoting asynchronous communication, and system openness as well—which is why they are widely adopted in the context of multi-agent systems (MAS), including mobile and intelligent agents [13]. To deal with the physical space as a first-class aspect, the extensions proposed in literature – such as Geo-Linda [14], TOTA

© Springer International Publishing AG 2017   121
C. Badica et al. (eds.), *Intelligent Distributed Computing X*,
Studies in Computational Intelligence 678, DOI 10.1007/978-3-319-48829-5_12

[10], LIME [15], $\sigma\tau$-Linda [21], Spatial ReSpecT [11] – enrich tuple spaces with an explicit physical location, given by the mobile hardware device on which they are meant to run.

In this paper we introduce a novel extension called Spatial Tuples, in which the basic information chunks – i.e., tuple themselves – have a *location* and *extension* in the physical space. A tuple space is then conceived as an *augmentation* of the physical reality, where tuples represent a living information layer mapped onto the physical environment. Assuming space as a first-class concept enhances generative communication with the ability to express different patterns of spatial coordination in quite a natural and effective way.

Here, the term "augmentation" is used to explicitly refer to (mobile) *augmented reality* [1, 5]: namely, assuming that *(i)* physical reality is enriched by digital information situated in some physical position, and that *(ii)* visually-augmented reality is perceived by human users by means of specific devices, such as smart-glasses, head-mounted-display, or even smartphones. Mobile-augmented and mixed reality are among the application domains where Spatial Tuples could be effectively used, in particular in scenarios where humans move inside the physical augmented environment, and have to coordinate with users and agents of any sort—both situated and non-situated ones. There, it can be used as a basic model to develop high-level forms of stigmergic coordination and cognitive stigmergy [16, 20] involving both humans and intelligent agents.

The *augmentation-based* view is a novel and original feature of Spatial Tuples with respect to the state-of-the-art, integrating concepts and views from different fields such as MAS, pervasive computing, and mixed/augmented reality—and thus, it strongly relates to the vision of *mirror words* as proposed in [17].

2   The Spatial Tuples Coordination Model and Language

In this section we introduce the main elements of the Spatial Tuples model and its coordination language by showing how it simply enhances the basic LINDA model (Subsection 2.1). We discuss its extension towards mobility (Subsection 2.2), then demonstrate how the model effectively tackles diverse sorts of spatial coordination issues through some simple examples (Subsection 2.3).

For the sake of simplicity, in the remainder of this section we adopt first-order logic notation for the specification of both tuples and tuple templates – as in logic-based coordination models like Shared Prolog [4] and TuCSoN / ReSpecT [12, 13] – where logic variables allow for partial specification of templates, and unification works as the matching mechanism. Accordingly, we adopt ground terms for regions (and locations), general logic term for region templates, and unification as their matching mechanism.

Since our focus here are the coordination issues, we do not explore the language issues involved in the description of location and regions—which are indeed essential for the application of Spatial Tuples to real-world scenarios, but at the same time orthogonal with respect to coordination issues. As a result, locations

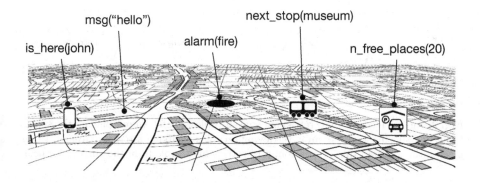

Fig. 1. Spatial tuples as information layer augmenting the physical reality.

and regions are simply represented by means of identifiers (logic terms), whereas their matching is based on their associated spatial properties.

2.1 Basics

Spatial tuples. Spatial Tuples deals first of all with spatial tuples. A *spatial tuple* is a tuple associated to a spatial information. Spatial information can be of any sort, such as: GPS positioning (latitude, longitude, altitude), administrative (via Zamboni 33, Bologna, Italy), organisational (Room 5 of the DISI building at the University of Bologna), etc.—whatever the case, it anyway associates the tuple to some *location* or *region* in the *physical space*.

As a result, a spatial tuple in principle *decorates* physical space, and can work as the basic mechanism for *augmenting reality* with information of any sort (see Fig. 1). Once a spatial tuple is associated to a region or location, its information can be thought of as describing properties of any sort that can be attributed to that specific portion of the physical space—thus implicitly adding observable properties to the physical space itself. By accessing the tuple with Spatial Tuples tuple-based mechanisms, the information can be observed by any sort of agent dealing with the specific physical space, so as to possibly behave accordingly.

In tuple space based models, the *communication language* is used to define the syntax of information content in tuples, as well as the matching mechanism between tuples and tuple templates. Besides, in Spatial Tuples a *space-description language* is introduced to specify the spatial information decorating the tuples and the corresponding matching mechanism. This spatial language is orthogonal to the communication language, and is meant to provide the basic ontology defining spatial concepts—such as locations, regions, places. To keep the description simple and general, in the remainder of the paper we will adopt an abstract spatial language, based on the intuitive concepts of point-wise *location* (no extension) and spatial *region* (with an extension).

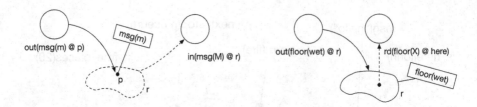

Fig. 2. Agents inserting and retrieving spatial tuples. *(Left)* An out placing the spatial tuple in a specific location; an in retrieving a tuple specifying a region. *(Right)* An out placing the spatial tuple in a region; a rd retrieving a tuple from the current location.

Spatial primitives. As in any tuple-based model, Spatial Tuples basic operations are out(t), rd(tt), in(tt), where t is a tuple, tt a tuple template—see Fig. 2 for a first intuition of how they are supposed to work.

Spatial tuples are emitted – that is, tuples are associated to a region or location r – by means of an out operation. The following invocation

$$out(t \ @ \ r)$$

emits spatial tuple t @ r—that is, tuple t associated to region (or location) r.

For instance, if p1 is a location and r2 a region, simple examples of out are

$$out(info(message) \ @ \ p1) \qquad out(floor(wet) \ @ \ r2)$$

respectively associating tuple info(message) to the physical position of location p1, and decorating region r2 with tuple floor(wet).

In tuple-based models, *getter* operations – in(tt), rd(tt) – look for tuples matching with *tuple template* tt. In Spatial Tuples they are extended with the ability to associatively explore the physical space, through the notion of *region template*: that is, the ability to describe either the region or the location in space also in a partial way, and to find tuples in the matching regions / locations.

Accordingly, the basic forms of getter operation in Spatial Tuples are

$$rd(tt \ @ \ rt) \qquad in(tt \ @ \ rt)$$

Both operations look for a tuple t matching tuple template tt in any location or region of space r matching spatial template rt: as in standard LINDA, in consumes the spatial tuple matching the template, whereas rd just returns it. Following the standard semantic of tuple-based models, getter primitives in Spatial Tuples are *(i) suspensive* – if no tuple matching tt is found in any region or location matching rt, then the operation is blocked until a matching tuple is made available somehow – and *(ii) non-deterministic*—if more than one tuple matching the tuple template is found in a region or location matching the spatial template, then one of them is returned non-deterministically. For instance,

$$rd(floor(F) \ @ \ p1)$$

looks for a tuple of the form floor(F) associated to location p1—if any tuple matching is found there, it is returned, otherwise the operation is suspended.

If no tuple of the form floor/1 occur in region r2, then rd(floor(F) @ r2) would block, to be subsequently resumed when an out(floor(dirty) @ p1) is performed, since p1 $\in$ r2 – so, location p1 spatially matches region (template) r2 – and tuple floor(dirty) matches tuple template floor(F) with substitution {F/dirty}—spatial tuple floor(dirty) @ p1 would be returned, in case.

The *predicative* and *bulk* extensions – represented by the primitives inp(tt @ rt), rdp(tt @ rt) and rd_all(tt @ rt, L), in_all(tt @ rt, L) – are defined analogously, and as such they do not require further discussion in this context.

2.2 Mobility

Spatial tuples can be associated to locations and regions either directly or indirectly. The above-defined notions and operations directly associated a spatial tuple t to a region / location r; in Spatial Tuples, however, a tuple can be associated to a system entity with either a position or an extension in the physical space. In the following, we will often refer to these entities as *situated components*.

So, if id identifies a component in a Spatial Tuples system (a software artefact, a physical device), and id is a situated component, a spatial tuple t can be associated to id by means of an out operation of the form

$$out(t \ @ \ id)$$

associating t to whichever is the region / location r where id is placed.

One of the main point here is that such an indirect association holds also when the region where id is located changes over time—that is, when id refers to a *mobile* entity: until it is removed by an in operation of some sort, t will be associated to whatever place in space id is situated during its life cycle.

By assuming, for the sake of simplicity, that logic terms are used also for identifiers, if car(1), car(2), and car(admin) are situated components in a Spatial Tuples system, operation

$$out(auth(level(max)) \ @ \ car(admin))$$

decorates component car(admin) with tuple auth(level(max)), which is associated indirectly to car(admin) position from now on, thus following any possible motion of car(admin).

As one may easily expect, also getter primitives are extended accordingly, allowing identifiers of situated components to be used as their targets. Then

rd(authorisation(level(L)) @ car(C))   in(authorisation(level(_)) @ car(admin))

respectively *(left)* reads the authorisation level L of any situated component car(C) (non-deterministically), and *(right)* removes any authorisation from car(admin), wherever it is located in space.

Also, an implicit form for getter primitives is available to agents actually *associated to situated components*: e.g., Spatial Tuples agents upon a mobile device. So, if agent a executes in component id, and id is in location p, operations of the form

$$out(t\ @\ here) \qquad out(t\ @\ me)$$

associate tuple t to current id location p. In the former case (here), the spatial tuple is permanently associated to p, even though id would subsequently move. In the latter case (me), the spatial tuple would be associated to p just for the time id stays in p, then following any motion of id.

In the same way, operations of the form

$$rd(tt\ @\ here) \qquad rd(tt\ @\ me)$$
$$in(tt\ @\ here) \qquad in(tt\ @\ me)$$

just return one tuple matching template tt if and when available at the current location of the situated component hosting the executing agent.

More articulated notations to specify / retrieve spatial tuples, situated components identifiers, and their positions in space altogether are available—but are not extensively discussed in the following, being not essential here.

2.3   Space-based Coordination in Spatial Tuples: Simple Examples

Few simple examples may help understanding how the basic Spatial Tuples model affects systems, effectively implementing different patterns of *spatial coordination*.

Breadcrumbs. A Spatial Tuples agent over a mobile device may simply implement the breadcrumbs pattern. For instance, agent hansel could hold a counter C, repeatedly increment it after a given period of time, and deposit a spatial tuple wasHere/2 just after, while moving, with an operation

$$out(wasHere(hansel,\ C)\ @\ here)$$

As a result, the trajectory of agent hansel could be observed by observing the spatial distribution of tuples wasHere/2, and possibly traced back.

Awareness. Spatial Tuples enables basic forms of spatial awareness. For instance, if a number of mobile devices is set to meet in region meeting, a meetingControl agent could check the arrival of all the expected devices by assuming that each of them would deposit a tuple through out(hereIAm(id) @ here), and getting all of them by repeatedly performing an

$$in(hereIAm(Device)\ @\ meeting)$$

until all the devices expected actually show up. It should be noted that while devices have to be situated components – so they can use the implicit form of the Spatial Tuples primitive –, meetingControl can be an agent of any sort, since the explicit form of the Spatial Tuples primitive does not require situation.

Situated knowledge sharing. Spatial Tuples makes it simple to design time-uncoupled communication that requires situated knowledge sharing. As a basic mechanism, in fact, Spatial Tuples allows tuples to be situated (in a location, in a region of space) with a single operation invocation. So, an agent is allowed to send a message to all agents located in some region indirectly, by representing the message as a spatial tuple associated to such a region.

For instance, if agent control_room in a rescue operation needs to warn all the rescuers about the presence of some injured people in some region, it could perform the invocation

$$out(warning(injured) @ region)$$

so that all rescuer agents situated in region, waiting for local warnings upon a

$$rd(warning(W) @ here)$$

would immediately receive the news.

Spatial synchronisation and mutual exclusion. For *spatial synchronisation* we mean synchronising actions of situated agents based on their position: e.g., an agent A would start some task as soon as an agent B arrives in some place.

Using again the meeting example above, an agent may decide to leave for a meeting only when any other agent reaches the meeting point through an invocation of the form:

$$rd(hereIAm(\_) @ meeting)$$

and waiting for the proper spatial tuple to be returned.

As a meaningful case, *spatial mutual exclusion* – for ruling the access of agents to some physical region – can be achieved quite simply in Spatial Tuples. E.g., in order to allow just one agent at a time in a region (mutex_region), to enter mutex_region an agent could be required to get a tuple lock situated there, to be released as soon as the agent exits the region, respectively through

$$in(lock @ mutex\_region) \qquad out(lock @ mutex\_region)$$

Spatial Dining Philosophers. A slight variation of the Dining Philosopher example can be easily adapted to showcase spatial coordination in Spatial Tuples. We do not focus here on deadlock issues, by simply assuming a trivial ticket-based solution where $N-1$ ticket @ table spatial tuples are initially in place, and each of the N philosophers enters the system with a in(ticket @ table).

As shown in Fig. 3, the chopsticks of a 4-Dining Philosophers problem are represented as spatial tuples chop situated in positions p1, p2, p3, and p4—chop@p1, chop@p2, chop@p3, chop@p4. All of them are placed on the table region (p1, p2, p3, p4 $\in$ table), and each of them is shared by two adjacent seats—so, for instance, p1 $\in$ seat4 $\cap$ seat1, p2 $\in$ seat1 $\cap$ seat2, and so on.

Thus, for a philosopher to eat from seat1, it is enough to move there, invoke in(ticket @ table), then perform twice the operation

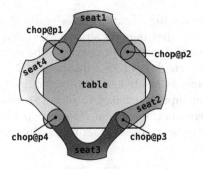

Fig. 3. Seats and table as regions for Spatial Dining Philosophers

$$in(chop @ seat1)$$

As a result, the philosopher would receive the two spatial tuples chop@p1 and chop@p2 – in any order, due to non-determinism – given that positions p1 and p2 spatially match with region template seat1.

Once both chop@p1 and chop@p2 have been obtained, the eating phase can start. As one may expect, releasing both tuples with

$$out(chop @ p1),\ out(chop @ p2)$$

and the ticket with an out(ticket @ table) ends the eating phase properly.

3   Discussion

A *proof-of-concept* prototype of Spatial Tuples was implemented on the TuCSoN middleware [13], which supports development of space-based models by: *(i)* providing geo-location services for mobile devices through its Android distribution; *(ii)* making it possible to customise the tuple matching function and the tuple language thanks to the tuProlog engine [6] used for tuple representation and matching; *(iii)* allowing for the extension of the set of coordination primitives available to coordinables by means of the ReSpecT language [12].

In the prototype, a Spatial Tuples layer – i.e., a tuple space storing spatial tuples – executes on a single host as a single TuCSoN node, acting as a main server for the whole spatial-augmentation layer. To ease the prototype engineering, a number of different ReSpecT tuple centres local to a single TuCSoN node are used to map different portions of the physical space.

The prototype is being tested against a *rescue* application scenario, where Spatial Tuples is used to coordinate *(i)* the rescuer assistant agents, exploiting wearable/mobile devices used by rescuers moving and acting inside the rescue field, *(ii)* the wounded-assistant agents, by relying on wearable devices placed (by rescuers) on the injured people found on the field, and *(iii)* control room agents, operating on the same server where the Spatial Tuples layer is deployed— the only not-situated entity in the system. In this scenario, many of the patterns of spatial coordination provided as examples in Subsection 2.3 are put to work: *breadcrumbs* are left by rescuers acting on the field for safety reasons, *awareness* is fundamental for the control room agents to monitor rescuers and wounded state, *knowledge sharing* is used to filter out unneeded information so as to avoid distracting the rescuers on field, *spatial synchronisation* and mutual exclusion enable efficient distribution of rescuers on the field, so as to avoid overlapping.

Beyond such an early prototype, a more distributed and scalable implementation can be devised by properly exploiting information about the space on which a tuple space is mapped, like in the case of spatial databases [8] and GIS (Geographical Information Systems) [2]. In the case of TuCSoN, for instance, the global spatial tuple space implementing the physical augmentation may be distributed on multiple nodes, where multiple ReSpecT tuple centres may run, each bound to some physical sub-region.

Additionally, geospatial properties can be further exploited to design a cloud-based version [22] of Spatial Tuples, running on top of a cloud computing infrastructure in charge of managing the tuple space, implemented, e.g., as a distributed hash map, as well as the reception, dispatching, and execution of the requests regarding spatial coordination primitives. This way, the Spatial Tuples model could be deployed within a mixed environment, composed by both *embodied* components, such as the agents executing on rescuers mobile devices, and *disembodied* ones, like the agents running within the cloud—e.g. those representing the control room.

Further issues still to be addressed will be subject of further investigation. For instance, the full exploitation of Spatial Tuples in real-world scenarios depends on the conception and design of a full-fledged space-description language, enabling expression of arbitrary matchmaking patterns among regions and locations of the physical space, suitably complementing the communication and the coordination languages used by tuple-based models.

4 Conclusion

This paper discusses a novel perspective on space-based coordination models for situated MAS, based on spatial tuples as chunks of information situated in the physical space, providing a digital layer augmenting the physical reality. The Spatial Tuples model integrates the power of generative communication within a framework where space is modelled as a first-class entity, thus allowing for different forms of spatial coordination, useful in application domains ranging from pervasive computing to mobile-augmented reality.

References

1. R. T. Azuma et al. A survey of augmented reality. *Presence*, 6(4):355–385, 1997.
2. N. Bartelme. Geographic information systems. In W. Kresse and M. D. Danko, editors, *Springer Handbook of Geographic Information*, pages 59–71. Springer, 2012.

3. J. Beal, D. Pianini, and M. Viroli. Aggregate programming for the Internet of Things. *IEEE Computer*, 48(9):22–30, Sept. 2015.
4. A. Brogi and P. Ciancarini. The concurrent language, Shared Prolog. *ACM Transactions on Programming Languages and Systems*, 13(1):99–123, Jan. 1991.
5. E. Costanza, A. Kunz, and M. Fjeld. Mixed reality: A survey. In D. Lalanne and J. Kohlas, editors, *Human Machine Interaction*, pages 47–68. Springer, 2009.
6. E. Denti, A. Omicini, and A. Ricci. tuProlog: A light-weight Prolog for Internet applications and infrastructures. In I. Ramakrishnan, editor, *Practical Aspects of Declarative Languages*, volume 1990 of *LNCS*, pages 184–198. Springer, 2001.
7. D. Gelernter. Generative communication in Linda. *ACM Transactions on Programming Languages and Systems*, 7(1):80–112, 1985.
8. R. H. Güting. An introduction to spatial database systems. *The VLDB Journal*, 3(4):357–399, 1994.
9. M. Hazas, J. Scott, and J. Krumm. Location-aware computing comes of age. *Computer*, 37(2):95–97, Feb 2004.
10. M. Mamei and F. Zambonelli. Programming pervasive and mobile computing applications: The TOTA approach. *ACM Transactions on Software Engineering Methodologies*, 18(4):1–56, 2009.
11. S. Mariani and A. Omicini. Space-aware coordination in ReSpecT. In M. Baldoni, C. Baroglio, F. Bergenti, and A. Garro, editors, *From Objects to Agents*, volume 1099 of *CEUR Workshop Proceedings*, pages 1–7, Turin, Italy, 2–3 Dec. 2013.
12. A. Omicini and E. Denti. From tuple spaces to tuple centres. *Science of Computer Programming*, 41(3):277–294, Nov. 2001.
13. A. Omicini and F. Zambonelli. Coordination for Internet application development. *Autonomous Agents and Multi-Agent Systems*, 2(3):251–269, Sept. 1999.
14. J. Pauty, P. Couderc, M. Banatre, and Y. Berbers. Geo-Linda: a geometry aware distributed tuple space. In *Advanced Information Networking and Applications*, pages 370–377, 2007. 21st International Conference (AINA '07), 21–23 May 2007, Niagara Falls, ON, CA. Proceedings.
15. G. P. Picco, A. L. Murphy, and G.-C. Roman. LIME: Linda meets mobility. In *The 1999 International Conference on Software Engineering (ICSE'99)*, pages 368–377. ACM, 1999. May 16-22, Los Angeles (CA), USA.
16. A. Ricci, A. Omicini, M. Viroli, L. Gardelli, and E. Oliva. Cognitive stigmergy: Towards a framework based on agents and artifacts. In D. Weyns, H. V. D. Parunak, and F. Michel, editors, *Environments for Multi-Agent Systems III*, volume 4389 of *LNCS*, pages 124–140. Springer, 2007.
17. A. Ricci, M. Piunti, L. Tummolini, and C. Castelfranchi. The mirror world: Preparing for mixed-reality living. *IEEE Pervasive Computing*, 14(2):60–63, 2015.
18. R. Scoble and S. Israel. *The Age of Context*. Patrick Brewster Press, Apr. 2014.
19. T. Starner. Wearable computers: no longer science fiction. *Pervasive Computing, IEEE*, 1(1):86–88, Jan 2002.
20. H. Van Dyke Parunak. A survey of environments and mechanisms for human-human stigmergy. In D. Weyns, H. Van Dyke Parunak, and F. Michel, editors, *Environments for Multi-Agent Systems II*, volume 3830 of *LNCS*, pages 163–186. Springer, 2006.
21. M. Viroli, D. Pianini, and J. Beal. Linda in space-time: an adaptive coordination model for mobile ad-hoc environments. In M. Sirjani, editor, *Coordination Models and Languages*, volume 7274 of *LNCS*, pages 212–229. Springer, 2012.
22. C. Yang, M. Goodchild, Q. Huang, D. Nebert, R. Raskin, Y. Xu, M. Bambacus, and D. Fay. Spatial cloud computing: how can the geospatial sciences use and help shape cloud computing? *International Journal of Digital Earth*, 4(4):305–329, 2011.

Exploring unknown environments with multi-modal locomotion swarm

Zedadra Ouarda[a], Jouandeau Nicolas[b], Seridi Hamid[a], Fortino Giancarlo[c]

[a] LabSTIC Laboratory, 8 may 1945 University
P.O.Box 401, 24000 Guelma, Algeria
[b] LIASD, Paris8 University, Saint Denis, France
[c] DIMES, Universita' della Calabria
Via P. Bucci, cubo 41c - 87036 - Rende (CS) - Italy

Abstract

Swarm robotics is focused on creating intelligent systems from large number of simple robots. The majority of nowadays robots are bound to operations within mono-modal locomotion (i.e. land, air or water). However, some animals have the capacity to alter their locomotion modalities to suit various terrains, operating at high levels of competence in a range of substrates. One of the most significant challenges in bio-inspired robotics is to determine how to use multi-modal locomotion to help robots perform a variety of tasks. In this paper, we investigate the use of multi-modal locomotion on a swarm of robots through a multi-target search algorithm inspired from the behavior of flying ants. Features of swarm intelligence such as distributivity, robustness and scalability are ensured by the proposed algorithm. Although the simplicity of movement policies of each agent, complex and efficient exploration is achieved at the team level.

Keywords: Swarm intelligence, Swarm robotics, Multi-target search, Random walk, Stigmergy, Multi-modal locomotion

1. Introduction

A search is defined as the action to look into or over carefully and thoroughly in an effort to find or discover something [1]. When agents lack information regarding targets, systematic searches become less effective and using random walk can enhance the chance of locating resources by increasing the chances of covering certain regions. In random strategies, the random walker (mobile robot or synthetic agent) returns to the same point many times before finally wandering away, which affects determinant parameters such as energy consumption [2], time and risks of malfunctions of agents. Stigmergy-based coordination allows very efficient distributed control and optimization. It has several other properties which are also essential to multi-robot systems, including robustness, scalability, adaptability and simplicity [3].

In [4] a cooperative and distributed coordination strategy called Inverse Ant System-Based Surveillance System (IAS-SS) is applied to exploration and

Preprint submitted to Elsevier   *May 29, 2016*

© Springer International Publishing AG 2017
C. Badica et al. (eds.), *Intelligent Distributed Computing X*,
Studies in Computational Intelligence 678, DOI 10.1007/978-3-319-48829-5_13

surveillance of unknown environments. It is a modified version of the artificial ant system, where the pheromone left has the property of repelling robots rather than attracting them. A guided probabilistic exploration strategy for unknown areas is presented in [5], it is based on stigmergic communication and combines the random walk movements and the stigmergic guidance. The paper [6], provides a simple foraging algorithm that works asynchronously with identical ants, based on marking visited grid points with pheromone. It lacks robustness to faults. Authors in [7], propose a swarm intelligence based algorithm for distribute search and collective clean up. In this algorithm, the map is divided into a set of distinct sub-areas and each sub-area is divided into some grid. Each robot decides individually based on its local information to which subarea it should move. A direct communication via WIFI model is used between robots and their neighbors. The paper [8], introduces the Ants Nearby Treasure Search (ANTS) problem, in which identical agents, initially placed at some central location, collectively search for a treasure in a two-dimensional plane without any communication. A survey of online algorithms for searching and exploration in the plane is given in [9].

Swarm robotics is the study of how a large number of simple physically embodied agents can be designed such that a desired collective behavior emerges from the local interaction among agents and between agents and the environment. The mono-modal locomotion has been the principal interest of swarm robotics [10] for so long period but also using heterogeneous robots with different locomotion has been investigated [11]. However, multi-modal locomotion seems to be very interesting in order to allow agents performing a variety of tasks adaptively in different environments. Swarm robotics with mono-modal locomotion remains an active research area whose promise remains to be demonstrated in an industrial setting. Swarm robotics with multi-modal locomotion constitutes a new orientation that can benefit from the developed applications and open the issue to the development of new coordination and cooperation strategies.

Flying Ant-like Searcher Algorithm (FASA), proposed in this paper, is a multi-target search algorithm. In order to avoid returning to the same place several times in a random walk search strategy we used stigmergic communication through pheromone to mark covered regions. Through simulations we observe that agents get stuck in covered regions when their number is high and only some of them can get out of the covered regions. Therefore we propose the flying behavior whenever the neighborhood is totally covered, then we use flying behavior to return to specific locations which we call *best positions*, memorized by the agent when its current cell has at least one neighbor not covered yet. These cells are considered best positions because they allow a gradual search from the starting point and the flying behavior of agents to such cells ensures that all previous cells will be covered before going far away from them. It is an algorithm with a low computational complexity and designed for agents (and so far for robots) with very simple low-range sensors and indirect communication known as stigmergy.

The rest of the paper is organized as follows: in section 2, we present the problem formulation. In section 3, we present the behavior of ants from which the proposed algorithm is inspired and then we give the finite state machine of our agents and the pseudo code of the proposed algorithm. In section 4, we present the scenarios and performance metrics used in simulations, after that, we present the obtained results and compare them with Random Walk (RW) and Stigmergic Random Walk (SRW). We finish with a conclusion and future perspectives in section 5.

2. Problem Formulation

In a collective multi-target search task, there are a lot of targets randomly distributed in an area. The agents (robots) should find as fast as possible the targets and, after that, remove them, if we deal with a cleanup task, or transport them to a nest, if we deal with a foraging task [2] [3]. In this work, we intend to design a search algorithm which allows a group of simple agents to locate a set of targets placed at random positions in the search space. The finish time of the collective search is when all targets have been found.

The basic concepts we use in the rest of the paper are defined as follow:

- *Environment*- A two dimensional finite grid E with NXM size. $E = E_{free} \cup E_{occupied}$, where $E_{occupied}$ denotes the subset of E containing the cell occupied by obstacles, targets or agents and $E_{free} = E_{covered} \cup E_{Ncovered}$ where $E_{covered}$ denotes the subset of E_{free} containing the covered cells (containing a pheromone) and $E_{Ncovered}$ denotes the subset of E_{free} containing the not yet covered cells. We define also E_{Best} as a subset of $E_{covered}$ containing the best positions stored by an agent. We denote a current cell C_c with coordinates $(x, y) \in E_{Best}$ if $\exists$ at least a neighbor cell C_n with coordinates $(x-1, y), (x+1, y), (x, y-1), (x, y+1) \in E_{Ncovered}$.

- *Target*- A set of static objects $T = t_1...t_n$, where n, the total number of targets is $>= 1$ and each t_i is placed at random positions in E.

- *Agent*- An Ant-like agent, which is capable of:

  1. Perceiving the four neighboring cells (detect the presence of pheromone, targets and obstacles);
  2. Depositing pheromone on current cell (to mark it as covered);
  3. Localizing itself;
  4. Moving and flying. Moving corresponds to one move from cell A to cell B with a distance of one step in one of the four directions up, down, left or right, while in flying the distance is $>= 1$;
  5. Memorizing the coordinates of best position cells;

- *Pheromone (ℙ)*- Chemical substance deposited by agents on visited cells to mark them as covered. It evaporates with time t.

3. Flying Ant-like Searcher Algorithm (FASA)

The FASA algorithm is a combination of random walk, stigmergic communication, and systematic search (using stored information). We provide the agents with the capacity to fly like flying ants. While workers of the Camponotus japonicus species [12] do not have wings, young female and male have them. They use them to fly away from their nest for mating and building their own colony. Figures 1(a), 1(b) and 1(c) represent the behavioral model of worker, male and female of Camponotus japonicus ants and Figure 2 represents the behavior of our Flying Ant-like Searcher agent which combines the behaviors of worker and male (or female) Camponotus japonicus ants.

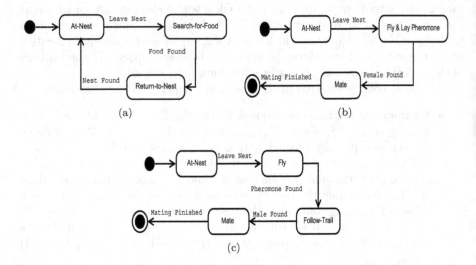

Figure 1: Behavioral model of: (a) worker, (b) male and (c) female Camponotus japonicus ants

Some particularities of FASA are: (1) agents explore gradually the sub-space in which they were initially deployed, (2) robustness to failure is provided as it works since a single agent is alive, (3) initial positions of agents, the geometry of the search space, complexity of obstacles do not influence the algorithm performances.

FASA consists of three steps:

1. Sets a temporization t to a random (value);
2. The agent repeats the following steps until $t = 0$:

   - Stores the coordinates of the current cell C_c in E_{Best}, if it has at least one neighbor not yet covered;

   - Deposits $\mathbb{P}$ on current cell C_c;

   - Moves to one of the four neighbors not yet covered;

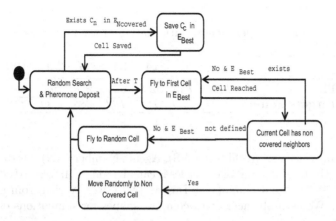

Figure 2: State transition diagram of FASA Agents

Algorithm 1: FASA

---

1 **while** *Targets not reached* **do**
2     $t \leftarrow$ random (value)
3     $E_{Best} \leftarrow \emptyset$
4   **while** $t \neq 0$ **do**
5   **if** $\exists\ C_n$ *(a neighbor of C_c)* $\in E_{Ncovered}$ **then**
6   push_front C_c in E_{Best}
7   Deposit $\mathbb{P}$ on C_c
8   Move randomly to $C_j \in E_{Ncovered}$
9        $t \leftarrow t - time\_step$
10   **if** $E_{Best} \neq \emptyset$ **then**
11   Fly to first element in E_{Best}
12   **while** *($E_{Best} \neq \emptyset$) and ($\nexists$ a neighbor of $C_{Best} \in E_{Ncovered}$)* **do**
13   Remove first element of E_{Best}
14   Fly to next $C_{Best} \in E_{Best}$
15   **while** *($C_c \in E_{covered}$) or (all neighbors of $C_c \in E_{covered}$)* **do**
16   fly randomly to new cell

---

3. Flies to the first best position in E_{Best}. If E_{Best} is void, keeps flying to random cells until finding a non covered cell or a cell having non covered neighbors and go to 1.

4. Performance Evaluation

4.1. Simulation Scenarios and Performance Metrics

Simulations are implemented and run on a Java-based simulation platform (Netlogo [13]). In all our experiments, targets were considered non-mobile. Ini-

Table 1: Experimental initialization of random value

	4	9	14	21	43	50	100
Time	1096	643	625	597	278	561	569
Targets Found	169	230	188	163	285	230	228

tial conditions such as: World Size (WS), Agents Number (AN), Obstacles Density (OD) – the total amount of obstacles is calculated by: $Amount of Obstacles = OD \times (\frac{worldsize}{4})$ – and Number of Targets (NT), were varied from a scenario to another. We used obstacles with complex shapes for simulations of obstacle environments.

We used two metrics to evaluate the performance of our algorithm. We compare it to random walk (RW) [14] and to Stigmergic Random Walk (SRW) where robots use stigmergic communication to avoid already visited cells when they lay pheromone trails [15]:

- Search Time: is the time in seconds needed to discover all the targets in the environment.

- Search Efficiency: is defined by 1:

$$Search_{eff} = \frac{Targets_{found}}{Targets_{total}} \times 100 \qquad (1)$$

where: $Search_{eff}$ denotes the percentage of found targets over the total number of targets, $Targets_{found}$ denotes the number of found targets during an elapsed time t and $Targets_{total}$ denotes the total number of targets placed at the environment.

We defined two scenarios where targets are placed randomly and agents start all from the center of the environment. The time t spent in random walk is fixed after experiments to random (x) with $x = 43$ (see Table 1). We varied value from 4 to 100 and then we recorded in the first scenario the average time required to find all the targets and in the second scenario the number of targets found. With Random (43), the time is the lowest and the number of targets found is the highest among the other values. Each simulation is performed 10 times, then the mean and standard deviation for each metric are computed:

- *Scenario 1:* to test the scalability of the algorithm, when increasing agents number. We fix the WS to 500 × 500 cells (Netlogo units), the NT to 40, the OD to 70% and we vary AN from 100 to 4000. We fix the time for each simulation to 240 sec and we report at the end the mean value of $Search_{eff}$.

- *Scenario 2:* to test the efficiency of search in larger environments. We fix AN to 300, NT to 40, OD to 70% and we vary WS from 200 × 200 cells to 1000 × 1000 cells.

4.2. Results and Discussion

Through the obtained results, FASA outperforms the two other protocols in both scenario 1 and 2. It is more efficient in locating targets and faster than the two others in searching the total number of targets in larger environments.

In scenario 1, the $Search_{eff}$ in FASA increases when increasing AN, it is about 48% with 100 agents and 100% over 300 agent. Also the search time decreases when increasing AN, from 225 sec (300 agents) to 95 sec (2000 agents) but over 4000 agents the search time starts at increasing (105 sec). FASA gives better results than SRW and RW algorithms. SRW's $Search_{eff}$ reaches 90% with 300 agents and then decreases till 28% with 4000 agents. RW gives the worst $Search_{eff}$ since there is no guidance in search, the $Search_{eff}$ increases slowly from 12% to 24% (with 100 to 800 agents respectively) over 800 agents it decreases till 7% with 4000 agents. Using large number of agents in SRW or RW causes agents to get stuck in already covered regions and agents keep turning in a closed covered region, while in FASA the flying behavior helps the agents to get out of the closed covered region and give them the chance to cover more regions and to find more targets (see Table 2 and Figure 3(a) for detailed results).

Table 2: $Search_{eff}$ of FASA, SRW and RW when increasing AN

	100	300	500	800	1000	2000	4000
FASA $Search_{eff}$	48%	100%	100%	100%	100%	100%	100%
FASA Search Time	240	225	164	113	106	95	105
SRW	40%	90%	25%	30%	32%	32%	28%
RW	12%	18%	18%	24%	23%	10%	7%

Table 3: Search Time of FASA, SRW and RW when increasing WS

	200x200	400x400	800x800	1000x1000
FASA	19	109	1213	2793
STD	1	10	83	111
SRW	201	1077	13554	26031
STD	46	354	1309	2761
RW	433	1686	20669	39941
STD	65	216	18661	37538

In scenario 2, search time increases in the three algorithms when increasing WS. FASA gives better results than SRW and RW. The search in SRW and RW becomes inefficient when world size is over 400 × 400 cells. In FASA the return to best positions by flying behavior results in gradual search over the whole

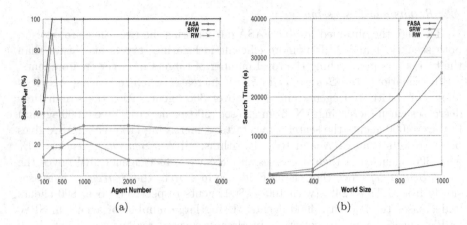

Figure 3: Simulation results of FASA, SRW and RW : (a) $Search_{eff}$ when increasing AN, (b) Search Time when increasing WS

search space even if there is some redundancy while searching (see Table 3 and Figure 3(b) for detailed results).

5. Conclusion

We presented in this paper a multi-target search algorithm called FASA. It tends to introduce guidance in search through pheromone, getting out from covered regions through the flying behavior and enabling a gradual search that ensures the completeness of the algorithm through the flying to best positions (stored while searching). Results obtained in comparison to random walk and pheromone guided random walk are promising. Parameters such as the random time of search and the distance of fly needs to be improved through simulations.

Stigmergic communication via pheromone has shown to efficiently coordinate a team of robots and to allow them to quickly explore a given area [16]. Simulations can support the analysis and improvement of pheromone-based algorithms before their real robotic implementation [5]. However, the implementation of the pheromone itself in real world constitutes a challenging technical issue. Several works proposed mechanisms to the real implementation of pheromone as: (i) physical marks using: *virtual marks* [16] or *RFID tags* [17], (ii) a model to be transmitted using wireless network [18], (iii) virtual pheromone transmitted using infrared communication [19], (iv) beacons where robots are used as pheromones [20]. Despite the proposed approaches, the implementation of pheromone is still in its early development stages and most of the works are available in research laboratories.

In order to test the applicability of the proposed algorithm, we intend to implement it in a robotic platform (ARGoS [21], Gazebo [22]) by also exploiting specific MAS-based methodologies [23] [24].

References

[1] V. Méndez, D. Campos, F. Bartumeus, Random search strategies, in: Stochastic foundations in movement ecology, Springer, 2014, pp. 177–205.

[2] O. Zedadra, H. Seridi, N. Jouandeau, G. Fortino, An energy-aware algorithm for large scale foraging systems, Scalable Computing: Practice and Experience 16 (4) (2016) 449–466.

[3] O. Zedadra, H. Seridi, N. Jouandeau, G. Fortino, A cooperative switching algorithm for multi-agent foraging, Engineering Applications of Artificial Intelligence 50 (2016) 302–319.

[4] M. F. R. Calvo, J. R. de Oliveira, R. A. F. Romero, Bio-inspired coordination of multiple robots systems and stigmergy mechanims to cooperative exploration and surveillance tasks, in: IEEE 5th International Conference on Cybernetics and Intelligent Systems (CIS), 2011, pp. 223–228.

[5] I. T. T. Kuyucu, K. Shimohara, Evolutionary optimization of pheromone-based stigmergic communication, in: Applications of Evolutionary Computation, Springer, 2012, pp. 63–72.

[6] C. Lenzen, T. Radeva, The power of pheromones in ant foraging, in: 1st Workshop on Biological Distributed Algorithms (BDA), 2013.

[7] A. L. D. Liu, X. Zhou, H. Guan, A swarm intelligence based algorithm for distribute search and collective cleanup, in: IEEE International Conference on Intelligent Computing and Intelligent Systems (ICIS), Vol. 2, IEEE, 2010, pp. 161–165.

[8] O. Feinerman, A. Korman, Z. Lotker, J. S. Sereni, Collaborative search on the plane without communication, in: Proceedings of the 2012 ACM symposium on Principles of distributed computing, ACM, 2012, pp. 77–86.

[9] S. K. Ghosh, R. Klein, Online algorithms for searching and exploration in the plane, Computer Science Review 4 (4) (2010) 189–201.

[10] B. Ranjbar-Sahraei, S. Alers, K. Tuyls, et al., Stico in action, in: Proceedings of the 2013 international conference on Autonomous agents and multi-agent systems, International Foundation for Autonomous Agents and Multiagent Systems, 2013, pp. 1403–1404.

[11] P. Pace, G. Aloi, G. Caliciuri, G. Fortino, A mission-oriented coordination framework for teams of mobile aerial and terrestrial smart objects, Mobile Networks and Applications (2016) 1–18.doi:10.1007/s11036-016-0726-4.

[12] T. Abe, On the behavior of the ant camponotus japonicus at the nuptial flight, Kontyu 41 (1973) 333–341.

[13] U. Wilensky, Netlogo. http://ccl.northwestern.edu/netlogo/, in: Center for Connected Learning and Computer-Based Modeling, Northwestern University, Evanston, IL, 1999.

[14] W. J. Bell, Searching behaviour: the behavioural ecology of finding resources, Springer Science & Business Media, 2012.

[15] L. Pitonakova, R. Crowder, S. Bullock, Understanding the role of recruitment in collective robot foraging, in: Proceedings of The Fourteenth International Conference on the Synthesis and Simulation of Living Systems (ALIFE), MIT Press, 2014, pp. 1477–1485.

[16] J. Svennebring, S. Koenig, Building terrain-covering ant robots: A feasibility study, Autonomous Robots 16 (3) (2004) 313–332.

[17] M. Mamei, F. Zambonelli, Spreading pheromones in everyday environments through rfid technology, in: 2nd IEEE Symposium on Swarm Intelligence, 2005, pp. 281–288.

[18] R. T. Vaughan, K. Støy, G. S. Sukhatme, M. J. Matarić, Blazing a trail: insect-inspired resource transportation by a robot team, in: Distributed autonomous robotic systems 4, Springer, 2000, pp. 111–120.

[19] D. W. Payton, M. J. Daily, B. Hoff, M. D. Howard, C. L. Lee, Pheromone robotics, in: Intelligent Systems and Smart Manufacturing, International Society for Optics and Photonics, 2001, pp. 67–75.

[20] E. J. Barth, A dynamic programming approach to robotic swarm navigation using relay markers, in: Proceedings of the American Control Conference, Vol. 6, IEEE, 2003, pp. 5264–5269.

[21] C. Pinciroli, V. Trianni, R. OGrady, G. Pini, A. Brutschy, M. Brambilla, N. Mathews, E. Ferrante, G. Di Caro, F. Ducatelle, et al., Argos: a modular, parallel, multi-engine simulator for multi-robot systems, Swarm intelligence 6 (4) (2012) 271–295.

[22] N. Koenig, J. Hsu, The many faces of simulation: Use cases for a general purpose simulator, in: Proc. of the ICRA, Vol. 13, 2013, pp. 10–11.

[23] G. Fortino, A. Garro, S. Mascillaro, W. Russo, Using event-driven lightweight dsc-based agents for mas modelling, International Journal of Agent-Oriented Software Engineering 4 (2) (2010) 113–140.

[24] G. Fortino, A. Garro, W. Russo, An integrated approach for the development and validation of multi-agent systems, International Journal of Computer Systems Science & Engineering 20 (4) (2005) 259–271.

Part V

Behavioral Analysis

GroupTrust: Finding Trust-based Group Structures in Social Communities

Antonello Comi, Lidia Fotia, Fabrizio Messina, Domenico Rosaci, and Giuseppe M.L. Sarné

Abstract Observing the features of the information actually stored in the Web, we can recognize that an important issue to be investigated is that of discovering relationships between groups of objects. In particular, a great interest is emerging on finding groups of objects mutually linked by reciprocal relationships of *trustworthiness*. In this paper, we propose a model to represent the case of trust-based groups of objects, and we present an algorithm for detecting trust associations in virtual communities in presence of these groups. Such an algorithm consists in determining particular sub-structures of the community, called *trust groups*, representing objects mutually connected by strong trust relationships. We technically formalize our idea and algorithm, and we present a complete example of how our approach works.

1 Introduction

Nowadays, in the most of social communities there is the possibility for the users to join with some *groups* to share common interests, discuss opinions and content of various type, organize events and other social activities. Groups catalyze the attention of large fractions of social communities and the overwhelming number of groups available on some community does not astonish us particularly: for example, Facebook and Twitter[4] users worldwide create more than 100,000 groups per

A.Comi, L.Fotia, D.Rosaci
DIIES, University Mediterranea of Reggio Calabria, Italy
e-mail: {antonello.comi,lidia.fotia,domenico.rosaci}@unirc.it

G.M.L.Sarné
DICEAM, University Mediterranea of Reggio Calabria, Italy – e-mail: sarne@unirc.it

F.Messina
Department of Mathematics and Computer Science, University of Catania, Viale A. Doria, 95125 Catania, Italy – e-mail: messina@dmi.unict.it

© Springer International Publishing AG 2017   143
C. Badica et al. (eds.), *Intelligent Distributed Computing X*,
Studies in Computational Intelligence 678, DOI 10.1007/978-3-319-48829-5_14

day. Groups do not contain only human users, but also several types of *objects* as e-Commerce products, e-Learning documents, etc.

Observing the characteristics of the information actually stored in the Web, as well as the features of the main Web applications, we can recognize that an important issue to be investigated is that of discovering relationships between groups of objects. In particular, a great interest is emerging on finding groups of objects mutually linked by reciprocal relationships of *trustworthiness*. The importance of trust in multi-agent communities has been widely recognized, and also the possibility to apply trust measures for selecting users and objects in a social network scenario has been investigated.

In most of Web activities, as in e-commerce, e-learning, e-government [2, 3], social networks and so on, objects are often grouped into collections, for representing data categories. This corresponds to the actual categorization of entities and resources to which data are associated as, for instance, groups of products in e-commerce or groups of users in social networks. These groups are often mutually related, or related to some single object. As widely recognized, many knowledge bases of interest today are best described as a linked collection of interrelated objects. Generally, in these groups objects are grouped based on some known common semantic relationship (e.g. "all the members of the group are interested to italian literature") and some formalisms have been proposed in Semantic Web literature to express these relationships, as RDF or OWL. However, beside the already known relationships that originated the groups, some other hidden associations could exist, explicitly not expressed but implicitly emerging between groups of objects. We faced this important issue in [10], by proposing an approach to detect the associations above. The main idea underlying that proposal was that of designing a model for groups of objects that can represent in a direct way both single entities and groups of entities, also allowing nesting of subgroups. This model describes the structure of a social community, and can be viewed as a collection of groups. In particular, a community can be viewed as a hierarchical structure, similar to that of a file system with files and directories, with the addition of the possibility to have relationships between the components of the community, i.e. the groups. A community can be also viewed as a generalization of a direct, labelled graph, in which the nodes can have a hierarchical structure.

In this paper, we adapt the above model to represent the case of groups of objects linked by trust relationhips, and we propose an algorithm for detecting trust associations in virtual communities in presence of groups of objects. Such an algorithm consists in determining particular sub-structures of the community, *trust groups*, that represent objects mutually connected by strong trust relationships. We technically formalize our idea and algorithm, and we present a complete example of how our approach works. The paper is structured as follows. In the next section, we deal with some related work. Section 3 provides technical details about our approach for finding groups connected by strong trust relationships, while Section 4 describes a concrete example of application of our approach to the case study of the Research Literature. Finally, in Section 5 we draw our conclusions and illustrate some possible future works.

2 Related Work

In many disciplines, there is a population of people which should be optimally divided into multiple groups based on certain attributes to collaboratively perform a particular task [13]. The problem becomes more complex when some other requirements are also added: homogeneity, heterogeneity or a mixture of teams, amount of consideration to the preferences of individuals, variability or invariability of group size, having moderators, aggregation or distribution of persons, overlapping level of teams, and so forth [5, 11]. In [7], the authors reveal how these problems can be mathematically formulated through a binary integer programming approach to construct an effective model which is solvable by exact methods in an acceptable time. Basu et al. [1] consider the problem of how to form groups such that the users in the formed groups are most satisfied with the suggested top-k recommendations. They assume that the recommendations will be generated according to one of the two group recommendation semantics, called Least Misery and Aggregate Voting. Rather than assuming groups are given, or rely on ad hoc group formation dynamics, their framework allows a strategic approach for forming groups of users in order to maximize satisfaction. Concerning the concept of trust, there exist in the literature several proposals. Sherchan et al. [12] present an important review of trust, in which they comprehensively examine trust definitions and measurements, from multiple fields including sociology, psychology, and computer science. Trust models [9] allow to exploit information derived by direct experiences and/or opinions of others to trust potential partners by means of a single measure [6]. Xia et al. [14] build a subjective trust management model AFStrust, which considers multiple factors including direct trust, recommendation trust, incentive function and active degree, and treats those factors based on the analytic hierarchy process (AHP) theory and the fuzzy logic rule. [8] describes how to build robust reputation systems using machine learning techniques, and defines a framework for translating a trust modeling problem into a learning problem.

3 Groups and Communities

In this section, we introduce the community model underlying our approach. The basic notion that we introduce is that of *community* , directly derived from that of community *defined* in [10], adapted to the case of a social community. The definition of community is based on the definition of *group*, representing a collection of objects hierarchically organized in sub-collections. The internal structure of a group is similar of that of a file system directory, containing either single objects or subgroups. However, differently from file system, where a directory and a file are different entities, the definition of a group is fully recursive. Also a single object is considered as a group (called *singleton*) and therefore a generic group is defined as a set of subgroups.

Definition 1 (Group). Let U be a set of objects. A *group* on U is either (*i*) a object (that we also call a *singleton group*) or (*ii*) a set of groups on U.

A main property we define on a group is that of *membership*. The members of a group f are all the groups that compose it, at each level of nesting. Then, we define the *memberset* of a group as the set of all its groups. As a particular case, we assume that a group has itself as its unique member.

Definition 2 (Memberset). Let U be a set of objects and let f be a group on U. The *memberset* of f, denoted by M_f, is a set of groups on U where $\forall g \in M_f$ either (*i*) $g \in f$ or (*ii*) $\exists k \in f$ such that $g \in M_k$. If $f \in U$, then $M_f = f$.

Property 1 **(Member of level *l*).** Let U be a set of objects and f be a group on U. We say that $g \in M_f$ is a *member of level* 0 of f if $g \in f$. We say that g is a *member of level l* of f if $\exists k \in M_f$ such that both $g \in k$ and k is a member of level $l - 1$ of f.

Based on the notion of group, we now define the notion of *community*. A community consists of a group and a set of labelled arcs that connects some of the members of the group. The structure of a community thus appears as an extension of a direct labelled graph, with the difference that the "nodes" of a community are the members of its group, that instead of necessarily representing a single object can have a more complex structure, with possible levels of nesting.

Definition 3 (Community). A *community* is a triple $\langle U, f, A \rangle$, where U is a set of objects, f is a group on U and A is a set of labelled arcs, such that each $a \in A$ is an ordered triple $\langle x, y, l \rangle$, where $x, y \in M_f$, and l is a label. We denote by $\emptyset$ the cardinality of U, by n the cardinality of M_f and by α the cardinality of A.

We define two types of relationships on a community, called *membership* and *trustlink*. A *membership* in a community is a relationship between two groups a and b members of f, such that b is member of a. A *trustlink* in a community is a relationship between two groups a and b members of f, such that there exists an arc oriented from a to b.

Definition 4 (Memberships). Let $F = \langle U, f, A \rangle$ be a community. The *memberships* of F, denoted by *memberships$_F$* is a relationship on $M_f \times M_f$ that contains all the ordered pairs $\langle a, b \rangle$, where $a, b \in M_f$ and $b \in M_a$.

Definition 5 (Trustlink). Let $F = \langle U, f, A \rangle$ be a community. The *trustlinks* of F, denoted by *trustlinks$_F$* is a relationship on $M_f \times M_f$ that contains all the ordered triple (a, b, l), where $a, b \in M_f$ and there exists an arc $a = \langle a, b, l \rangle \in A$.

Note that between two groups a and b, members of a community F, only an instance (a, b) can exist in *memberships$_F$*, while many instances $\langle a, b, l \rangle$ can exist in *trustlinks$_F$*.

A community has got a number of members equal to the cardinality of M_f, i.e. n. In its turn, each of these member, say m, has got a number of members equal to $|M_m|$.

Therefore, the cardinality of *memberships$_F$*, that we denote by $\psi = \sum_{m \in f} |M_m|$.

It is possible to provide a representation of a community $F = \langle U, f, A \rangle$ by using a labelled directed graph that contains, for each object u of F a correspondent node $u*$, and for each member m of f two nodes m^{arr} and m^{dep}, called *arrival node* and *departure node*, respectively, where if m is an object then $m^{arr} = m^{dep} = m*$. For each arc a of A, oriented from a group x to a group y, a corresponding arc, called *trustlink arc*, directed from x^{dep} to y^{arr} and labelled with the same label of a, is inserted in the graph-representation. Moreover, each node m^{arr} is linked by a fictitious arc, called *membership arc*, with the arrival node of each element of m, to represent the fact that each arc incoming in m^{arr} has to be joined with each element of m. Analogously, each departure node of the groups contained in m is linked by another membership arc with the node m^{dep} to represent the fact that each element of m is joined with each arc outcoming from m^{dep}. A conventional label *MEMBER* is applied to all the membership arcs.

Definition 6 (Graph-representation). Let $F = \langle U, f, A \rangle$ be a community. The graph-representation of F, denoted by G_F, is the labelled directed graph $\langle N_F, A_F \rangle$, where (*i*) for each object $u \in U$ a correspondent node $u*$ is inserted in N_F and for each group $m \in \mathcal{M}_f$, two nodes m^{arr} and m^{dep} are inserted in N_F such that $m^{arr} = m^{dep} = m*$ if $m \in U$ and (*ii*) for each pair of groups $x, y \in \mathcal{M}_f$ such that $y \in \mathcal{M}_x$, both an arc $\langle x^{arr}, y^{arr}, MEMBER \rangle$ and an arc $\langle y^{dep}, x^{dep}, MEMBER \rangle$ are inserted in A_F (these two arcs are called *membership arcs* and (*iii*) for each arc $\langle x, y, l \rangle \in A$, an arc $\langle x^{dep}, y^{arr}, l \rangle$ is inserted in A_F.

It is possible to define for a community the notion of *path* between two groups.

Definition 7 (Path). Let $F = \langle U, f, A \rangle$ be a community and x, y be two members of f. We suppose to have two functions, namely $ini : A \rightarrow \mathcal{M}_f$ and $fin : A \rightarrow \mathcal{M}_f$ such that for each arc $e = \langle x, y, l \rangle \in A$, we have $ini(e) = x$ and $fin(e) = y$. A *path* between x and y is a sequence of arcs $a_1, a_2, .., a_k \in A$, such that $x \in \mathcal{M}_{ini(a_1)}$, $y \in \mathcal{M}_{fin(a_k)}$ and $ini(a_{i+1}) \in \mathcal{M}_{fin(a_i)}$, $\forall i = 1, 2, k-1$.

Since the arcs of a community are labelled, and the label of an arc represents an information characterizing the relationship between the groups linked by the arc, we introduce the notion of $\mathcal{F}$-*relevant path*, that is a path whose arcs present values of the labels satisfying a given boolean function $\mathcal{F}$.

Property 2 ($\mathcal{F}$-relevance of a path). Let $F = \langle U, f, A \rangle$ be a community, $p = a_1, a_2, .., a_k \in A$ be a path in F and $\mathcal{F}$ be a boolean function that accepts as input a path and returns either *true* or *false*. The path p is called $\mathcal{F}$-*relevant* if $\mathcal{F}(p) = true$.

We can also define other properties for a community, analogously to similar properties of a graph, as that of *connection* between two groups and that of *strongly connected component*.

Property 3 (Connection). Let $F = \langle U, f, A \rangle$ be a community and $x, y \in \mathcal{M}_f$ be two members of f. Moreover, let $\mathcal{F}$ be a boolean function accepting as input a path of F. We say that x and y are *trust-connected* (resp. $\mathcal{F}$-*relevant trust connected*) if there exists at least a path (resp. a $\mathcal{F}$-*relevant path*) in F from x to y. Each member $x \in \mathcal{M}_f$ is connected (resp. $\mathcal{F}$-*relevant trust connected*) with itself.

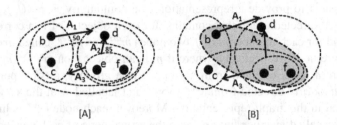

Fig. 1 [A] a community and [B] its unique strongly trust connected component (in grey)

Definition 8 (Strongly ($\mathcal{F}$-relevant) trust connected components). Let $F = \langle U, f, A \rangle$ be a community and let $\mathcal{F}$ be a boolean function accepting as input a path of F. A strongly trust connected component (resp. a strongly $\mathcal{F}$-relevant trust connected component) of F is a group f^* on U^* where (i) $U^* \subseteq U$ and (ii) $m \in \mathcal{M}_f \; \forall m \in \mathcal{M}_{f^*}$ and (iii) for each oriented pair of groups (a, b), where $a, b \in f^*$, we have that a and b are connected (resp. F-relevant connected) in F.

Example 1 ($\mathcal{F}$-relevant strongly trust connected components). Consider the community of Figure 1-[A], and the boolean function $\mathcal{F}(p)$ accepting as input a path p and returning *true* if all the arcs composing p have a label value greater than 30. It is easy to see that its unique $\mathcal{F}$-relevant strongly connected component is the group $\{b, \{e, f\}\}$, highlighted by a grey ellipse in Figure 1-[B], since b is connected to $\{e, f\}$ being linked by the arc $A1$ (having label value equal to 50) to $\{d, \{e, f\}\}$ and $\{e, f\}$ is connected to b being linked by the arc $A3$ (having label value equal to 60) to $\{b, c\}$. For better understanding this result, it is sufficient to apply the standard algorithm for finding the strong connected components to the graph representation of this community.

4 Discovering trust groups in a community: The case of Research Literature

It is interesting to point out that the group $\{b, \{e, f\}\}$ determined as the unique $\mathcal{F}$-relevant strongly trust connected component in the community of Figure 1-[A] is not a member of the community. In other words, determining such a group as the result of finding the $\mathcal{F}$-relevant strongly trust connected components of the community has led us to discover a structure embedded in the community, not explicitly "declared" as a part of the community, and that naturally emerges in consequence of the arc-relationships and member-relationships existing in the community itself. We call such a type of group a *trust group* (T-groups, for short).

Definition 9 (Trust groups). Let $F = \langle U, f, A \rangle$ be a community and $\mathcal{F}$ be a boolean function accepting as input a path of F. A *trust group* (T-group) on $\mathcal{F}$ of F is a $\mathcal{F}$-relevant strong trust connected component c of F such that $c \notin \mathcal{M}_f$.

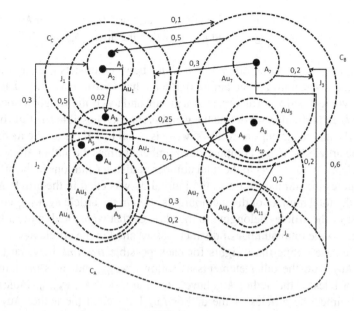

Fig. 2 A community associated to the scientific database

Finding trust groups in a community F allows us to discover new, potentially interesting information about the community from a trust viewpoint. Indeed, if we assume that all the elements of a group are trust-related, a trust group f determined into F represents a trust relationship between its elements, and thus it can be considered as a new "semantic frame" individuated in the environment represented by the community F. We will see below some examples of how such a kind of information can be usefully exploited in the context of Literature Research.

Now, we explain how it is possible to use groups to generate the recommendations for the researchers inserted into a scientific database. Modeling a scientific database as a community can make possible to capture the hidden relationships between objects and groups. As an example, we propose to model each *class of journal* by a group, each journal by a subgroup, each author by a sub-subgroup and each article by an object (see Figure 2). In particular, we suppose that in this scientific database, there are three main classes called C_A, C_B and C_C, dealing with three categories of articles. The expert evaluators classify the journals according to the relevance of the scientific contribution (i.e., C_A denotes the highest value). Moreover, we suppose that the group C_A is partitioned in two subgroup of journals called J_2 and J_4, the group C_B contains only the subgroup J_3 and, finally, the group C_C is composed by the subgroup J_1. Furthermore, we presume that each subgroup is associated to multiple authors. As shown in Figure 2, this last statement implies that the authors can belong to several subgroups as they may have made a scientific contribution to more journal that belong to different classes. For example, the sub-subgroup Au_2 represents an author who has made a contribution both to the journal J_2 and to the journal J_1, belonging to C_A and C_C, respectively. Moreover, an author can write one or more articles along with other authors. In the case of Au_2, we have three objects (i.e. articles) A_3, A_4 and A_5. Note that the article A_4 was also written by the author Au_3.

In this context, we define different *trustlinks* between objects (i.e., articles) and groups (i.e., classes, journals and authors). Recall that a *trustlink* is defined as a tuple $\langle X, Y, w_{XY} \rangle$ where X and Y are an object or a group, and w_{XY} is the weight associated to the *trustlink*. In particular, we define the weights for each possible *trustlink* between the object A_X and the groups as follows: $w_{A_X A_Y}$ is equal to 1 if the article A_Y has been cited by the article A_X; $w_{A_X Au_Y}$ computes the number of times the author Au_Y was cited in the article A_X, with respect to the total number of distinct citations in the article A_X; $w_{A_X J_Y}$ computes the number of times the journal J_Y was cited in the article A_X, with respect to the total number of distinct journals cited in the article A_X; $w_{A_X C_Y}$ computes the number of the articles belonging to the class C_Y that were cited in the article A_X, with respect to the total number of distinct classes cited in the article A_X.

Moreover, we define the weights for each possible *trustlink* between the sub-subgroup Au_X with the other elements as follows: $w_{Au_X A_Y}$ indicates the number of times the articles of the author Au_X have cited the article A_Y; $w_{Au_X Au_Y}$ indicates the number of times the articles of the author Au_X have cited the author Au_Y; $w_{Au_X J_Y}$ indicates the number of times the articles of the author Au_X have cited the journal J_Y; $w_{Au_X C_Y}$ indicates the number of times the articles of the author Au_X have cited the articles belonging to the class C_Y. These weights have been normalized in the interval $[0 \cdots 1]$, by dividing by the total number of distinct citations in the articles of the author Au_X.

Again, we define the weights for each possible *trustlink* between the subgroup J_X with the other elements as follows: $w_{J_X A_Y}$ indicates the number of times the articles of the journal J_X have cited the article A_Y; $w_{J_X Au_Y}$ indicates the number of times the articles of the journal J_X have cited the author Au_Y; $w_{J_X J_Y}$ indicates the number of times the articles of the journal J_X have cited the articles of the journal J_Y; $w_{J_X C_Y}$ indicates the number of times the articles of the journal J_X have cited the articles belonging to the class C_Y. These weights have been normalized in the interval $[0 \cdots 1]$, by dividing by the total number of distinct citations in the articles of the journal J_X.

Finally, we define the weights for each possible *trustlink* between the group C_X with the other elements as follows: $w_{C_X A_Y}$ indicates the number of times the articles of the class C_X have cited the article A_Y; $w_{C_X Au_Y}$ indicates the number of times the articles of the class C_X have cited the author Au_Y; $w_{C_X J_Y}$ indicates the number of times the articles of the class C_X have cited the articles belonging to the journal J_Y; $w_{C_X C_Y}$ indicates the number of times the articles of the class C_X have cited the articles belonging to the class C_Y. These weights have been normalized in the interval $[0 \cdots 1]$, by dividing by the total number of distinct citations in the articles of the class C_X.

Note that the normalization operation was carried out to make comparable the *trustlinks'* weights. In Figure 2, it is shown some *trustlinks* inside the community. Now, in order to exploit such representation to generate recommendation for an author regarding the research groups that could join or journals where might publish new articles, we could compute the trust groups on $\mathcal{F}$ of the community, where

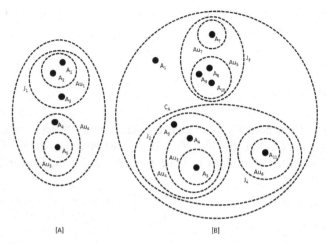

Fig. 3 Two naturally emerging groups.

$\mathcal{F}$ is a boolean function that accepts as input a path and returns *true* if the labels of the path have a value greater than 0,4. This means to consider the objects and the groups as semantically connected only if each link composing the path between them assumes the value at least equal to 0,5. The most important trust groups are represented in Figure 3. For instance, we can suggest to the author Au_1 to collaborate with the authors Au_3 and Au_4 because the journal J_1 in which the author Au_1 publishes the articles contains several citations to these authors (see the new group in Figure3-[A]). Moreover, we can suggest to the authors Au_1, Au_5 and Au_7 to publish the articles in the class C_A because the article A_1 and the articles A_7 to A_{10} are closely related to the issues of the journals J_2 and J_4. Finally, we can suggest the formation of a new research group to which belong the authors Au_1 and Au_3 to Au_7 because they work on common issues (see the new group in Figure3-[B]).

5 Conclusion

The importance of discovering trust relationships between members of social communities that have a group-structure is assuming particular relevance, since finding these relationships implies to determine sub-communities of members that could effectively interact with each other. In this paper, we have proposed to model such a kind of scenario by a semantic Web model proposed in the past, that has been adapted to represent trust relationship, and we have introduced an approach to discover groups of members not explicitly connected by trust relationships but that can be found by exploring the so obtained representation of the community. In this paper, we limited ourselves to introduce and formalize the idea, and we present an example of how the presented approach can found relevant trust connected groups. Our ongoing research is currently devoted to apply the approach to real situations, in which the advantages and limitations introduced by our proposal can be quantitatively and effectively evaluated.

Acknowledgment

This work has been partially supported by the Program "Programma Operativo Nazionale Ricerca e Competitività" 2007-2013, Distretto Tecnologico CyberSecurity funded by the Italian Ministry of Education, University and Research.

References

1. S. Basu Roy, L. V. Lakshmanan, and R. Liu. From group recommendations to group formation. In *Proceedings of the 2015 ACM SIGMOD International Conference on Management of Data*, pages 1603–1616. ACM, 2015.
2. F. Buccafurri, L. Fotia, and G. Lax. Allowing non-identifying information disclosure in citizen opinion evaluation. In *Technology-Enabled Innovation for Democracy, Government and Governance*, pages 241–254. Springer, 2013.
3. F. Buccafurri, L. Fotia, and G. Lax. Allowing privacy-preserving analysis of social network likes. In *Privacy, Security and Trust (PST), 2013 Eleventh Annual International Conference on*, pages 36–43. IEEE, 2013.
4. F. Buccafurri, L. Fotia, and G. Lax. Social signature: Signing by tweeting. In *Electronic Government and the Information Systems Perspective*, pages 1–14. Springer, 2014.
5. A. Comi, L. Fotia, F. Messina, G. Pappalardo, D. Rosaci, and G. M. Sarné. Forming homogeneous classes for e-learning in a social network scenario. In *Intelligent Distributed Computing IX*, pages 131–141. Springer, 2016.
6. A. Comi, L. Fotia, F. Messina, D. Rosaci, and G. M. Sarnè. A qos-aware, trust-based aggregation model for grid federations. In *On the Move to Meaningful Internet Systems: OTM 2014 Conferences*, pages 277–294. Springer, 2014.
7. A. A. Kardan and H. Sadeghi. An efficacious dynamic mathematical modelling approach for creation of best collaborative groups. *Mathematical and Computer Modelling of Dynamical Systems*, 22(1):39–53, 2016.
8. X. Liu, A. Datta, and E.-P. Lim. *Computational Trust Models and Machine Learning*. CRC Press, 2014.
9. E. Majd and V. Balakrishnan. A trust model for recommender agent systems. *Soft Computing*, pages 1–17, 2016.
10. D. Rosaci. Finding semantic associations in hierarchically structured groups of web data. *Formal Aspects of Computing*, 27(5-6):867–884, 2015.
11. I. K. Savvas and M. T. Kechadi. Mining on the cloud: K-means with mapreduce. In *2nd Int. Conference on Cloud Computing and Service Sciences, CLOSER*, pages 413–418, 2012.
12. W. Sherchan, S. Nepal, and C. Paris. A survey of trust in social networks. *ACM Computing Surveys (CSUR)*, 45(4):47, 2013.
13. M. Wessner and H.-R. Pfister. Group formation in computer-supported collaborative learning. In *Proceedings of the 2001 international ACM SIGGROUP conference on supporting group work*, pages 24–31. ACM, 2001.
14. H. Xia, Z. Jia, Z. Ju, X. Li, and Y. Zhu. A subjective trust management model with multiple decision factors for manet based on ahp and fuzzy logic rules. In *Green Computing and Communications (GreenCom), 2011 IEEE/ACM International Conference on*, pages 124–130. IEEE, 2011.

Non-intrusive Monitoring of Attentional Behavior in Teams

Davide Carneiro[1,3], Dalila Durães[2], Javier Bajo[2], and Paulo Novais[3]

[1] CIICESI, ESTGF, Polytechnic Institute of Porto, Portugal
[2] Department of Artificial Intelligence, Technical University of Madrid, Madrid, Spain
[3] Algoritmi Center/Department of Informatics, University of Minho, Braga, Portugal

Abstract. Attention is a very important cognitive and behavioral process, by means of which an individual is able to focus on a single aspect of information, while ignoring others. In a time in which we are drawn in notifications, beeps, vibrations and blinking messages, the ability to focus becomes increasingly important. This is true in many different domains, from the workplace to the classroom. In this paper we present a non-intrusive distributed system for monitoring attention in teams of people. It is especially suited for teams working at the computer. The presented system is able to provide real-time information about each individual as well as information about the team. It can be very useful for team managers to identify potentially distracting events or individuals, as well as to detect the onset of mental fatigue or boredom, which significantly influence attention. In the overall, this tool may prove very useful for team managers to implement better human resources management strategies.

Keywords: Attentional Behavior, Non-intrusive, Monitoring, Distributed Computing

1 Introduction

Attention is a very complex process through which one individual is able to continuously analyze a spectrum of stimuli and, in a sufficiently short amount of time, chose one to focus on [1]. In most of us, which can only focus on a very reduced group of stimuli at a time, this implies ignoring other perceivable stimuli and information.

Research on attention involves nowadays many fields, including education, psychology, neuroscience, cognitive neuroscience and neuropsychology. For this reason, many different views and theories on attention can be found. One of the most frequent ones is the so-called *Attention Economics*, which treats human attention as a scarce commodity or resource, which we must use wisely in order to attain our goals [2].

Individuals who have difficulties in focusing their attention can see the performance of other high-level cognitive processes negatively affected, such as learning or decision making. In extreme cases, such as in Attention-Deficit/Hyperactivity

© Springer International Publishing AG 2017
C. Badica et al. (eds.), *Intelligent Distributed Computing X,*
Studies in Computational Intelligence 678, DOI 10.1007/978-3-319-48829-5_15

Disorder (ADHD), this may have a significant negative impact in the development and function of the individual [3].

Although these aspects have always existed, in the last years we have witnessed the increasing of distracting stimuli, which make this topic even a more important one. Nowadays we have to deal with constant notifications from our e-mail, our social networks, our messaging applications, advertisements and so on. We live immerse in beeps, vibrations, notifications and blinking icons, which constantly call for our attention and distract us [4]. Even if we return immediately to our task, the fact that we had to consciously evaluate the stimuli to decide that it is not important at the moment had already a toll on our brain, making it spend resources [2, 5].

This is especially worrying in young children, who nowadays have a facilitated access to computers, mobile phones and tablets, with their games and engaging applications. For them it is so easy to get distracted by these stimuli, making learning less efficient, and more frustrating, affecting negatively their development [6].

In this paper we present a distributed system for monitoring attention in teams of people, in line with the vision of Intelligent Environments[7]. It is especially suited for people working with computers and can be interesting for domains such as the workplace or the classroom. It constantly analyzes the behavior of the user while interacting with the computer and, together with knowledge about the task, it is able to temporally classify attention.

This work may be very interesting for team managers to assess the level of attention of their teams, identifying potentially distracting events, hours or individuals. Moreover, distraction often appears when the individual is fatigued, bored or not motivated. This tool can thus be an important indicator of the team, allowing the manager to act accordingly at an individual or group level. In the overall, this tool will support the implementation of better human resources management strategies.

1.1 Previous Work

Part of the framework presented in this paper was implemented in previous work. This first version focused on the analysis of people's interaction patterns with the computer, including features such as mouse velocity or acceleration, click duration, typing speed or rhythm, among others. For a complete list of features as well as the process of their acquisition and extraction, please see [8].

While this early work focused on the detection of stress [8] and mental fatigue [9] from the analysis of Human-Computer Interaction, we also found out that people tend to interact differently with different applications and in different contexts. For example, and although both tasks involve typing, people tend to type differently if they are in a messaging application or in a word processing application [10].

For relatively simple domains in which people interact with a reduced and known set of applications, we achieved correct classification rates of around 90%

[11]. However, in more complex scenarios and, especially, if there are no limits on the possible applications, the accuracy of the classification tends to decrease.

The present work adds a new feature to this previously existing framework, by providing a precise measure of attention based not on the key typing or mouse movement patterns but on the actual application usage and switching patterns. It thus constitutes a much more precise and reliable mechanism for attention monitoring, while maintaining all the advantages of the existing system: non-intrusive, lightweight and transparent.

2 Architecture

The architecture of the developed system (Figure 1) is divided in three major parts. The lower-level is composed by the devices that generate the raw data (e.g. computers). These devices store the raw data locally in SQLite databases, until it is sychnronized with the web servers in the cloud, which happens at regular intervals.

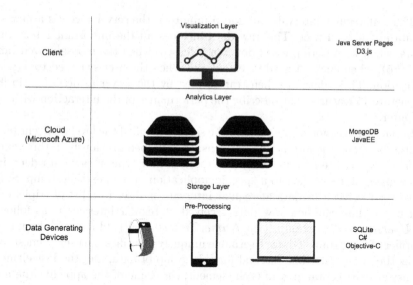

Fig. 1. Architecture of the developed system.

The main element in the middle layer is a MongoDB database, which is half way between relational and non-relational systems. It is actually more than a data storage engine, as it also provides native data processing tools: MapReduce and the Aggregation pipeline. Both the aggregation pipeline and map-reduce can operate on a shared collection (partitioned over many machines, horizontal scaling). These are powerful tools for performing analytics and statistical analysis in real-time, which are useful for ad-hoc querying, pre-aggregated reports, and more. MongoDB provides a rich set of aggregation operations that process data records and return computed results; using these operations in the data layer simplifies application code and limits resource requirements.

Finally, the visualization layer (top-most layer) is developed as a web app on Java technology and uses the D3 library for graphics and diagrams. It includes a set of intuitive data visualization tools to facilitate decision making and human resources management, with a focus on individual and group performance real time analytics.

2.1 Data Acquisition

As previously mentioned, the early version of this system acquired data describing the interaction of the user with the computer in terms of the mouse and the keyboard [9]. Data acquisition is supported by an application installed in the data-generating devices (the computers of the users). It is thus a distributed data collection system, which has now been extended to acquire a new type of data about each user: the application that the user is interacting with, at each time.

Thus, at regular intervals (around 5 minutes), the server receives a new set of data about each user. This raw data includes all the important interaction events: when keys were pressed down or released, when the mouse moved (and where to), when clicks started or ended and when the user switched to a given application. These data are then transformed by the server as described in [8], to generate 15 features that describe the performance of the interaction with the computer.

In the early version of the system, the server classified the level of attention of the user based on the interaction patterns, as described previously in this paper. With this work, the server now has access to a new type of raw data denoting that a given user switched to a specific application at a given timestamp. Since the server receives this data at regular intervals, it receives a list of triplets for each user. Thus, the new raw data received at regular intervals is as follows: $(Id, Username, Timestamp, [(AppName, Timestamp)])$. Id denotes the unique identifier of this group of data, Username uniquely identifies the user, Timestamp marks the end of the time interval for this group of data, and the last element is a list of pairs containing, in each element, the name of the app that the user switched to and the timestamp in which this happened.

2.2 Feature Extraction

When the server receives the data described in the previous subsection, it transforms it so its features can be extracted. Specifically, it goes through the list of pairs and computes the time during which each window was active (Algorithm 1). There are often cases in which the user does not change applications for a large amount of time. In these cases, which are represented by a pair with an empty AppName, the time is added to the last known AppName (since this means that the user is still interacting with it). An example of the output of this process is depicted in Figure 2.

Application	End Date	Use Time
record	Thu 3 Mar 2016 14:45:41 GMT	2.95345 min
Creative Cloud	Thu 3 Mar 2016 14:45:41 GMT	0.122 s
PRIMAVERA Windows Services Taskbar Icon	Thu 3 Mar 2016 14:45:44 GMT	3.299 s
Novo separador - Google Chrome	Thu 3 Mar 2016 14:45:55 GMT	10.265 s
MEO Cloud - Google Chrome	Thu 3 Mar 2016 14:46:28 GMT	32.849 s
Adobe Reader Updater	Thu 3 Mar 2016 14:49:43 GMT	3.25032 min
Espaão de armazenamento do perfil	Thu 3 Mar 2016 14:49:45 GMT	2.659 s
MEO Cloud - Home - Google Chrome	Thu 3 Mar 2016 14:50:01 GMT	15.576 s
Little Mix - Secret Love Song (Official Video) ft. Jason Derulo - YouTube - Google Chrome	Thu 3 Mar 2016 14:51:14 GMT	1.21132 min
Microsoft Access	Thu 3 Mar 2016 14:51:22 GMT	8.047 s
MEO Cloud - Home - Google Chrome	Thu 3 Mar 2016 14:51:37 GMT	15.137 s
Microsoft Access - M2_TESTE02_Alunos : Base de Dados (Access 2007 - 2010)	Thu 3 Mar 2016 14:51:39 GMT	2.079 s

Fig. 2. Sequence of applications used by a specific student, with the date in which the student switched to other application and the time spent interacting with it.

The next step is to compute the level of attention of the user (Algorithm, 2. To do this we measure the amount of time, in each interval, that the user spent interacting with work-related applications. The algorithm thus needs knowledge about the domain in order to classify each application as belonging or not to the set of work-related applications. This knowledge is provided by the team administrator and is encoded in the form of regular expressions. The team administrator uses a graphical interface to set up rules such as "starts with Microsoft" or "Contains word Adobe", which are then translated to regular expressions that are used by the algorithm to determine which applications are work-related.

Whenever an application that does not match any of the known rules for the specific domain is found, the application name is saved so that the team manager can later decide if a new rule should or should not be created for it. By default, applications that are not considered work-related are marked as "others" and count negatively towards the quantification of attention. Attention is calculated at regular intervals, as configured by the team manager (e.g. five minutes). The output of the algorithm can be visualized in Figure 3.

With this approach, several interesting functionalities can be implemented that provide valuable information to improve the team manager's decision making processes. Examples of such functionalities are described in Section 3.

Data:
p - A list of pairs of the type (AppName, Timestamp)
ft - the finishing time of the task
Result: durations - A list of triplets of the type (AppName, Timestamp,
          Duration)
durations ← [];
i ← 0;
while $i < Size(p)$ **do**
    task ← $p_{i,1}$;
    time ← $p_{i,2}$;
    i++;
    while $i < Length(p)$ and $StringLength(p_{i,1}) = 0$ **do**
        | i++;
    end
    if $i = Length(p)$ **then**
        | AppendTo(durations, task, ft, ft - time);
    else
        | AppendTo(durations, task, $p_{i,1}$, $p_{i,1}$-time)
    end
end

Algorithm 1: Creating triplets with the duration and timestamp of each application.

Data:
t - A list of triplets of the type (AppName, Timestamp, Duration)
st - The starting time of the task
inter - The interval to update attention
set - the set of regular expressions
Result: attention - A list of triplets of the type (timestamp, attention%,
          others%)
attention ← [];
i ← 0;
work ← 0;
others ← 0;
time ← st;
while $i < Size(t)$ **do**
    if $isWork(t_{i,1},\ set)$ **then**
        | work ← work + $t_{i,3}$;
    else
        | others ← others + $t_{i,3}$;
    end
    if $t_{i,2} > time + inter$ **then**
        AppendTo(attention, $t_{i,2}$, $work * 100/(work + others)$,
        $others * 100/(work + others)$));
        work ← 0;
        others ← 0;
        time ← $t_{i,2}$;
    end
end

Algorithm 2: Creating triplets at regular intervals with the timestamp and the quantification of attention.

Date	% Work	% Others
📅 Thu 3 Mar 2016 14:49:43 GMT	88.8875	11.1125
📅 Thu 3 Mar 2016 14:54:49 GMT	43.9485	56.0515
📅 Thu 3 Mar 2016 14:59:49 GMT	86.9204	13.0796
📅 Thu 3 Mar 2016 15:04:51 GMT	74.5224	25.4776
📅 Thu 3 Mar 2016 15:10:25 GMT	99.5259	0.47408
📅 Thu 3 Mar 2016 15:15:28 GMT	92.614	7.38601
📅 Thu 3 Mar 2016 15:20:41 GMT	99.3591	0.640935

Fig. 3. Detailed evolution of attention of a specific student.

3 Validation

As previously mentioned, a system with these characteristics may prove useful in very different domains, including organizational, academic or any environment in which people operate computers. In order to validate the proposed system, we have been using it for the past months in the Caldas das Taipas High School, located in northern Portugal. In the Portuguese academic context, this system gains increased relevance as current policies move towards the creation of larger classes, which make it increasingly difficult for the teacher to individually address each student. In this section we show several tools supported by this system that, when at the disposal of teachers may allow to:

- Decide, in real-time, which students to focus on according to their level of attention;
- Evaluate, *a posteriori*, which contents are more prone to generate distraction, providing a chance for improvement;
- Identify, in real-time, fluctuations in attention, improving decision-making concerning aspects such as when to make breaks or when to dismiss the class.

To validate this system we are following several cohorts of students during their academic activities. This extensive data collection process will allow to assess the influence of attention on aspects such as: breaks, time of the day, class contents, class objectives, among others. In this section, as an example, we briefly analyzed the data collected for the same cohort of students (10N) in two different classes: a regular one and an assessment one. Apart from the aims, the conditions were the same: the same cohort of students working on similar tasks, which required the use of Microsoft Access and Adobe Acrobat Reader.

Figure 4 allows the teacher to analyze, at the end of the class, the amount of time that each student spent at the computer (Task Duration) as well as the amount (and percentage) of time that each student devoted to work and to other activities. This is important for the teacher to perform a self-evaluation of how the class took place.

Student	Task Duration	Work	Work %	Others	Others %
T2240001	85.0084 min	0	0. per second	5099.54 s	100.
T2240003	90.0093 min	3765.67 s	69.7298	1634.7 s	30.2702
T2240004	90.0065 min	3065.46 s	56.7655	2334.75 s	43.2345
T2240005	90.0103 min	3075.89 s	56.9518	2324.98 s	43.0482
T2240006	90.0083 min	3271.33 s	60.5955	2127.3 s	39.4045
T2240007	95.0146 min	3524.17 s	61.8121	2177.25 s	38.1879
T2240008	70.0049 min	3516.83 s	83.727	683.522 s	16.273
T2240009	70.0069 min	3720.91 s	88.5797	479.725 s	11.4203
T2240010	80.0047 min	2965.08 s	61.7682	1835.26 s	38.2318
T2240011	75.0066 min	3767.64 s	83.7166	732.827 s	16.2834
T2240013	70.0048 min	3171.96 s	75.5176	1028.33 s	24.4824
T2240014	75.006 min	3159.89 s	70.2127	1340.57 s	29.7873

Fig. 4. The amount of time that each student spent interacting with the computer and the amount of actual work versus the amount of time spent interacting with other applications.

If necessary, the teacher may also click on a student to analyze the temporal evolution of the attention for that specific student, in a given class. Figure 5 shows the evolution of attention for three specific students during the class.

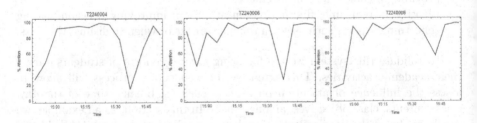

Fig. 5. Temporal evolution of attention in three different students of the same class.

The teacher may also find it very important to assess, in real-time or *a posteriori*, the evolution of attention of the whole class. To this end, he may select which cohorts to compare and in which classes. Figure 6 shows the global evolution of the attention of cohort 10N, in a regular class (a) and in an assessment class (b). This visual representation is constructed by combining data from all the students and computing a running average. Finally, several summarization techniques are also available, with the aim of providing the teacher with simple and intuitive insights into the data. As an example, Figure 6 (c) shows the distribution of the values of attention in cohort 10N, in a regular class and in an assessment one.

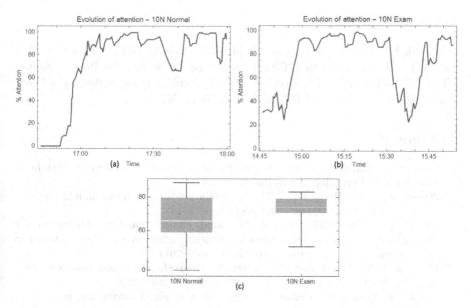

Fig. 6. Temporal evolution of attention in the same cohort of students (10N) in two different situations: regular class (a) and assessment class (b). Comparing the overall attention of both groups (c).

4 Limitations and Future Work

The work developed so far resulted in a very useful system for team managers to monitor, in real-time, the level of attention of their team. However, a limitation was also identified, being user T2240001 (depicted in Figure 4) an example. This is a user that, according to the collected data, has 0% of attention throughout the 85 minutes of the task. In fact, what happened is that the user opened an application that is not work-related and did not interact anymore with the computer until the end of the task. Similarly, if the user opens a work-related application and does not interact with the computer after that, the user's attention will be classified as 100% when he is most likely not even at the computer. These cases must, evidently, be pointed out.

To address this limitation, in future work we will implement a tighter integration between the previous and the new work. Specifically, in previous work we implemented the monitoring of the interaction of the users with the keyboard and the mouse. It is thus possible to know all the actions that each user performed either with the mouse or the keyboard, and at what time. We will thus generate a new feature that will quantify the level of activity of each user throughout time. This new feature will allow a more contextualized analysis of attention, improving the performance of its classification and quantification.

Acknowledgments

This work has been supported by COMPETE: POCI-01-0145-FEDER-007043 and FCT – Fundação para a Ciência e Tecnologia within the Project Scope: UID/CEC/00319/2013. The work of Davide Carneiro is supported by a Post-Doctoral Grant by FCT (SFRH/BPD/109070/2015).

References

1. Estes, W.K.: Handbook of Learning and Cognitive Processes (Volume 4): Attention and Memory. Psychology Press (2014)
2. Davenport, T.H., Beck, J.C.: The attention economy: Understanding the new currency of business. Harvard Business Press (2013)
3. ATTENTION-DEFICIT, S.O., et al.: Adhd: clinical practice guideline for the diagnosis, evaluation, and treatment of attention-deficit/hyperactivity disorder in children and adolescents. Pediatrics (2011) peds–2011
4. McBride, D.L.: Distraction of clinicians by smartphones in hospitals: a concept analysis. Journal of advanced nursing 71(9) (2015) 2020–2030
5. Simola, J., Hyönä, J., Kuisma, J.: Perception of visual advertising in different media: from attention to distraction, persuasion, preference and memory. Frontiers Media SA (2015)
6. Gottlieb, J.: Attention, learning, and the value of information. Neuron 76(2) (2012) 281–295
7. Augusto, J.C., Callaghan, V., Cook, D., Kameas, A., Satoh, I.: Intelligent environments: a manifesto. Human-Centric Computing and Information Sciences 3(1) (2013) 1–18
8. Carneiro, D., Novais, P., Pêgo, J.M., Sousa, N., Neves, J.: Using mouse dynamics to assess stress during online exams. In: Hybrid Artificial Intelligent Systems. Springer (2015) 345–356
9. Pimenta, A., Carneiro, D., Novais, P., Neves, J.: Monitoring mental fatigue through the analysis of keyboard and mouse interaction patterns. In: Hybrid Artificial Intelligent Systems. Springer (2013) 222–231
10. Pimenta, A., Carneiro, D., Novais, P., Neves, J.: Detection of distraction and fatigue in groups through the analysis of interaction patterns with computers. In: Intelligent Distributed Computing VIII. Springer (2015) 29–39
11. Carneiro, D., Pimenta, A., Gonçalves, S., Neves, J., Novais, P.: Monitoring and improving performance in human–computer interaction. Concurrency and Computation: Practice and Experience (2015)

A Speculative Computation Approach for Conflict Styles Assessment with Incomplete Information

Marco Gomes, Tiago Oliveira and Paulo Novais

Department of Informatics, University of Minho, Braga, Portugal
{marcogomes,toliveira,pjon}@di.uminho.pt

Abstract. This paper analyses a way to cope with incomplete information, namely information regarding the conflict style used by parties. This analysis is important because it enables us to develop a more accurate and informed conflict style classification method to promote better strategies. To develop this proposal, an experiment using a combination of Bayesian Networks with Speculative Computation is depicted. Thus, in this work, was firstly identified and applied a set of methods for classifying conflict styles with incomplete information; secondly, the approach was validating opposing data collected from a web-based negotiation game. From the experiment outcomes, we can concluded that it is possible to cope with incomplete information by producing valid conflict style default values and, particularly, to anticipate competing postures through the dynamic generation of recommendations for a conflict manager. The findings suggest that this approach is suitable for handling incomplete information in this context and can be applied in a viable and feasible way.

Keywords: Intelligent System, Speculative Computation, Conflict Styles

1 Introduction

In a conflict resolution process, parties may assess the same situation in different ways, and, as such, respond differently. It is, therefore, necessary to understand the conflict (response) styles of the parties involved so as to manage conflicts properly. Another important step in this process is the one of the actual negotiation between the parts. A step in which they have to come up with concrete and valid outcomes, which may be rejected or probably changed by the other part [1]. A basic ingredient of this process is the correct anticipation of the parties' actions. In this sense, and whatever specific conflict style is used to solve or to manage the conflict, both parties may be attempting resolution of the conflict by acting based on their individual perception of the situation. However, a perception based on the faulty or incomplete information could undermine the resolution process and can lead to erroneous conclusions and therefore to an inexact evaluation of consequences. And this, surely, can be problematic for parties or a conflict manager who do not have a complete picture of each other's behaviours.

Therefore, a cornerstone of this approach is the idea that if a conflict manager, in an intelligent conflict support system, has the knowledge to be able to identify parties' conflict styles he will promote better strategies and, consequently, lead to better

© Springer International Publishing AG 2017

C. Badica et al. (eds.), *Intelligent Distributed Computing X*,
Studies in Computational Intelligence 678, DOI 10.1007/978-3-319-48829-5_16

solutions. Our work is, hence, devoted to investigations on how to cope with incomplete information when analysing the parties' conflict styles. Additionally, this kind of analysis is critical to enable us to develop a more accurate and informed conflict style classification method. Furthermore, this work describes the design of a speculative computation framework to cope with incomplete information regarding the parties' conflict style assessment. In general terms, this approach can be seen as a tentative procedure when complete information about conflict style is not obtained. Although the following terms may be associated with different meanings, the terms "conflict style" and "negotiation style", in this section and throughout this study, are used interchangeably. The remainder of this paper is structured as follows. Section 2 presents a brief review of a conceptual framework to facilitate information gathering in a municipal conflict support system. Further, the means by which the conflict style is classified in a digital environment are depicted. A case scenario is presented. In Sections 3 and 4, we describe our approach to improving the classification of conflict styles with incomplete information in a generic municipal decision-making process. The final section details the main conclusions drawn from this study.

2 Municipal Intelligent Conflict Support System

The proposed system is a conceptual framework which can be implemented as a computer-based support system to assist municipal decision-making when handling conflict. The system consists of components underpinning an intelligent environment that could be sensitive and responsive to both parties of a conflict management process. Pragmatically speaking, the system was designed to sense conflict context, acquire it and then make reasoning on the acquired context and thus acting on the parties' behalf. On the one hand the system builds up a consented information database of each and can link that subsequently with the proper individual performance within the conflict process that is monitored by the system. On the other hand, while the user conscientiously interacts with the system and takes his/her decisions and actions, a parallel and transparent process takes place in which contextual and behavioural information is sent in a synchronized way to the conflict support platform. Upon converting the sensory information into useful data, the platform allows conflicting manager for a contextualized analysis of the user's data. It is out of the scope to explain in detail each component of the system. Despite that, regarding the main aim of this work, the way how conflict styles are assessed in a digital environment is depicted as follows.

2.1 Conflict Style Assessment Model

Conflicts can develop in stages and consequently may involve many different responses as the conflict proceeds. People involved develop various strategies, solutions or behaviours, to deal with the conflict. The style of managing a conflict that each one has must be seen as having a preponderant role in the outcome of a conflict resolution process, especially on those in which parties interact directly (e.g., negotiation, mediation). To classify the conflict style, the proposals must be analysed, namely regarding their utility. In that sense, in each stage of the negotiation the parties' proposals are analysed according to their utility value and a range of possible outcomes defined by the values of the Worst Alternative to a Negotiated Agreement (WATNA) and Best Alternative to a Negotiated Agreement (BATNA) of each party. This approach uses a mathematical model [2], which classifies a party's conflict style considering the range of possible outcomes, the values of WATNA and BATNA as boundaries, and the utility of the proposal. Regarding that utility, it quantifies how good a given outcome is for a party, it is acceptable to argue that a competing party will generally propose solutions that maximize its own utility in expense of that of the other party (the utility of the proposal is higher than the WATNA of the other party), whereas, for example, a compromising party will most likely search for solutions in an intermediary region (the utility of the proposal falls within the range of the zone of possible agreement, the range of overlapped outcomes that would benefit both parties). Essentially, we were able to classify the personal conflict style of a party by constantly analysing the utility of the proposals created. Once the styles are identified, strategies can be implemented that aim to improve the success rate of procedures for resolution and conflict management.

2.2 Municipal Service Delivery - Proof of Concept

The proof of concept was chosen to demonstrate the applicability of the concepts underpinning this study. Employing this method, it aimed to prove the viability and feasibility of innovative concepts through prototypes and demonstrations of features. In this case, the proof was designed purely to demonstrate the functionality of Speculative Computation approach to deal with incomplete information regarding the classification of conflict styles. The broader aim is to experiment a path of research in which innovative methods to manage conflict could be integrated into a generic municipal decision-making process. This first attempt at performing something that might be real-world usable comprises the definition of a scenario, a simulation environment and analysis of the outcomes. As stated before, for the purpose of simulating a conflict situation in real-life municipal environments a previous developed web-based simulation was adapted to this purpose. It was adapted to enable test participants having a conflict experience induced by the presence of an Ambient Intelligence system. In that sense, a game was designed to simulate a scenario in which a municipality needs to perform a service contract (an agreement between a municipality and a service provider) to guarantee the repairs and maintenance of municipal equipment. Each party has to achieve a desired result in the negotiation, in other words, the negotiation outcome was to be a win/win situation for both sides. The game starts with the application randomly giving one of the predetermined roles to each party. The instructions to win the game were to negotiate a fair deal and make sure that the party in question did not miss the deal. Each party's instructions were clearly presented, visible to them through the application interfaces. To classify the conflict style, the proposals must be analysed regarding their utility. In each stage of the negotiation, the parties' proposals are analysed according to their utility value and a range of possible outcomes defined by the values of the worst alternative to a negotiated agreement (WATNA) and best alternative to a negotiated agreement (BATNA) of each party. Regarding the conflict style analysis, the ZOPA (Zone Of Possible Agreement) was bounded by the BATNA (10000 Euro) and WATNA (12000 Euro) values. The range of possible agreement was 2000, but the parties were not aware of this detail. The participants of the proposed experiment were volunteers socially connected with our lab members. Forty individuals participated, both female and male,aged between 22 and 53. During the experiments, the information about the user's context was provided through a monitoring framework, which is customized to collect and treat the interaction data. The participants played the web-based game through computers that allowed the analysis of the described features. The obtained results comprise data about negotiation experiments carried out in a controlled environment in our lab. The results will be analysed in following sections.

3   A Probabilistic Model for Conflict Style Assessment

The key issue to using a Speculative Computation approach to cope with incomplete information in a negotiation, namely information regarding the negotiation style used by participants, is to be able to generate a likely set of default parameters. The same is to say, to be able to predict the negotiation styles of participants in each round. The intuition here is that the conflict style exhibited by a participant in a round will affect the conflict style used by the very same participant in the following round. A predictive model capable of this has to produce the most likely values for a set of parameters, in this case for the negotiation style and establish dependence relationships between the negotiation styles in different rounds. Networks (BNs), for their set of characteristics [3], fulfil these requirements.

3.1   Bayesian Networks

BNs are graphical representations of statistical dependences and independences between variables [3]. They provide a network structure and a probability distribution that are easily interpreted by humans and machines. The problem at hand goes beyond the typical classification and regression problems. It is, instead, a density estimation problem where the objective is to find the most likely collective state of information for the negotiation styles in successive rounds of negotiation. As such, BNs fit the problem description better than other machine learning models [3]. Formally, a BN is an acyclic directed graph $G = (V(G), A(G))$ with a set of vertices $V(G) = \{V_1, ..., V_n\}$ where each vertex $Vi \in V(G)$ represents a discrete stochastic variable, and a set of arcs $A(G) \subset V(G)XV(G)$ where each arc $(V_i, V_j) \in A(G)$ represents a statistical dependence. A BN defines a joint probability distribution Pr that may be factorized according to Equation 1, where π is associated with the set of variables that denote the vertex parents of V_i.

$$Pr(V_i, ..., V_n) = \prod_{i=1}^{n} Pr(V_i | \pi(V_i))$$ (1)

3.2   Building the Predictive Model

Since previous analyses [4] revealed that the role of a participant significantly influences his negotiation style, Sellers and Buyers should be modelled in different networks. Understanding the dependence relationships between negotiation styles in different rounds is possible by performing BN structure learning on the data gathered from the experiments. Some of the most well known BN structure learning algorithms include: the *hill-climbing* (*hc*) search [5], the *grow-shrink* (*gs*) [5], the *max-min hill-climbing* (*mmhc*) and *restriction maximization* (*rsmax2*) [6], and the *chow-liu* algorithm. Using the above-mentioned algorithms, the goal is to obtain the network structure that best describes the dependences between negotiation styles in different rounds, based on the available data. Different scores can be used to assess the fitness of the network. In the present work, the Akaike Information Criterion (AIC) [7] was the selected metric. It is defined as in Equation 2.

$$AIC = logL(X_1, ..., X_n) - d \qquad (2)$$

The *logL* component corresponds to an assessment of goodness of fit, of how well a model fits the data. The *d* component is a penalty that is an increasing function of the number of estimated parameters and is used to avoid overfitting. Given Equation 2, the goal is to maximize the AIC. With a BN structure and the corresponding probability distribution, it is possible to retrieve the values for the variables that maximize the distribution, given available evidence. This is modelled as a *Maximum a Posteriori* estimation (MAP), defined as in Equation 3 [8].

$$MAP_\theta = \underset{\theta}{argmax} \, Pr(\theta|e) \qquad (3)$$

The θ component represents the goal variables for which the estimation is calculated and *e* represents the available evidence. This will be the mechanism by which the default values for negotiation styles in successive rounds of negotiation will be produced. These values will be used in a Speculative Computation framework that models interventions in the negotiation in order to influence the participants to adopt *collaborating* negotiation styles.

3.3 Determining Default Values for Conflict Styles

The above-mentioned experiments produced data regarding 20 negotiation games, which means that there were data for 20 Sellers and 20 Buyers. Out of all the games, the minimum number of rounds was two, the maximum was 12, and the most frequent number of rounds was three (observed in nine games). It was also possible to observe that 95% of the games had three or more rounds. Given this analysis it was decided that the model would feature predictions up to three rounds of negotiation. As such, two data sets were produced, one for Sellers and another for Buyers. Each data set had three columns representing the first three rounds, filled in with the negotiation styles exhibited by 19 participants for each data set. Only the records of the game with two negotiation rounds were excluded.The structure learning algorithms mentioned in the previous section were applied to the two data sets, producing the AIC scores shown in Table 1.

At the same time, a network structure was modelled based on the underlying intuition for the work, i.e, that the negotiation style exhibited in a round, by both Sellers and Buyers, would be probabilistically dependent on the negotiation style exhibited in

Table 1: Results of the structure learning and structure assessment processes.

Network Model	Seller AIC score	Buyer AIC score
hc	-38.749	-78.361
gs	-40.233	-78.361
mmhc	-40.233	-78.361
rsmax2	-40.233	-78.361
chow-liu	-40.795	-79.990
expert	-40.795	-79.990

the previous round. Let *Round1*,*Round2*, and *Round3* be variables that represent the negotiation styles in rounds 1, 2, and 3, this supposition can be translated into the network structure. The network was called *expert*. In Table 1, it is possible to observe that the AIC scores of the network structures were, overall, very low. This was most likely due to the reduced size of the sample used to learn them. However, given the complex data collecting process described above, the experiments would have had to be carried out on a much larger scale in order to produce better results. That being said, based on the available scores, the best network structure for the Seller role was the one resulting from the *hc* algorithm. The network is shown in Figure 1a and its score was even higher than the *expert* network structure, thus contradicting the initial intuition. It has a diverging arc disposition and the negotiation styles of *Round2* and *Round3* are shown to be conditionally independent given *Round1*. Throughout the rounds, it is possible to observe that Sellers remain competitive, but the probability of their assuming a *collaborating* negotiation style does increase. Regarding the networks for the Buyer role, the *hc*, *gs*, *mmhc* and *rsmax2* algorithms produced the same type of structure, which was also, coincidentally, the highest scoring network. It is presented in Figure 1b. Again, the score was higher, although slightly, than that of the *expert* network. It was only possible to establish a probabilistic dependence relationship between the negotiation styles of *Round1* and *Round2*. *Round3* was considered to be independent from the other two. Be that as it may, the network reveals that Buyers usually start with a *competing* negotiation style, but they are the ones to respond to the lack of dynamism in the negotiation by changing the negotiation style towards *accommodating* in the third round, thus unlocking a possible impasse.

    With the networks of Figure 1 and using MAP queries it is possible to obtain the most likely negotiation styles for the three rounds before the negotiation starts, or midway, for the remaining rounds, after a round is over. Supposing one is at the beginning of the negotiation, no round has taken place yet, and there is no information about previous negotiation styles, it would be possible to determine the most likely set of negotiation styles of a Seller for each round. Performing the MAP query with no evidence on the network of Figure 1a would result in assuming, by default, that *Round1*, *Round2*, and *Round3* would have the value *competing*, with a probability of 0.563. Now, supposing the first round is over and the Seller adopted a *collaborating* style instead, the MAP query to determine the most likely styles in the next two rounds would have *Round1* = *collaborating* as evidence. This query would produce two sets of predictions.

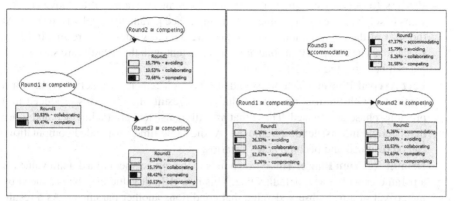

(a) Selected network structure for Sellers   (b) Selected network structure for Buyers

Fig. 1: Bayesian network structures.

The first would be *Round2 = collaborating* and *Round3 = collaborating*, and the second would be *Round2 = collaborating* and *Round3 = acommodating*. Both sets would have a probability of 0.500. These predictions are used in the Speculative Computation framework modelling the interventions of the conflict manager in the negotiation.

4 Speculative Computation Approach to Cope with Incomplete Information

The part taken by the conflict manager in the negotiation is that of trying to bring it to a satisfactory conclusion to both parties, to make them collaborate. As such, it has to anticipate the negotiation style used by the participants in each round, fill in this incomplete information, and recommend changes to the behaviours of the participants. This is modelled using Speculative Computation.

4.1 Speculative Computation Framework

Speculative Computation is a logic programming framework that structures reasoning with incomplete information, based on the use of default values. Its complete semantics and procedures are presented in [9, 10]. The framework has two principal components:

- **Default set (Δ):** This is a set containing default values for the parameters that serve as premises in the inference process;
- **Logic Program ($\mathscr{P}$):** This is a set of rules that allow the drawing of conclusions based on specific parameters;

The normal execution of the framework includes providing values for the parameters and traversing the rules in $\mathscr{P}$ until a conclusion is reached. The execution is structured in three phases:

- **Process Reduction Phase:** This corresponds to the derivation of tentative conclusions based on the existing values for the inference parameters (defaults or not). At the start of the computation all the values are default values. This results in likely scenarios. The computations that are not consistent with the defaults are suspended, but can be activated in a later phase;
- **Fact Arrival Phase:** This corresponds to the replacement of a default value for a parameter with the true observed value. As a result, the scenarios produced in a previous phase are revised. Those that are still consistent are maintained and those that become inconsistent are removed. At the same time, suspended computations can be resumed and become new scenarios;
- **Default Revision Phase:** This corresponds to the replacement of a default value for a parameter with a new default value. This new default value may be produced by the arrival of a true observed value that conditions another parameter. As a result, the scenarios produced in a previous phase are revised.

Speculative Computation fits systems that have a well-defined logic. It can serve as an interface between a predictive model and the rules that guide the enactment of system functions. Its advantage is in the clear definition of the procedures to manage the derived conclusions in a logic framework, according to the ever-changing state of the information.

4.2 Generating Recommendations for the Conflict Manager

The main procedures occurring in the conflict manager assume the form of Speculative Computation phases. The worflow of the conflict manager is depicted in Figure 2.

The set of rules that allow the inference of the recommendations the conflict manager may provide to Sellers or Buyers, based on their negotiation styles, is represented in the $\mathscr{P}$ component of the framework. At the beginning of the negotiation, when there is no information about the negotiation styles, a MAP query, such as the one performed in Section 3.3, provides the initial default set Δ. As shown in Figure 2, through a process reduction phase this would result in tentative scenarios representing possible recommendations of the conflict manager. For instance, in the example given above, it was ascertained that the most likely negotiation style of a Seller in the first three rounds would be *competing*. *Competing* would be then used in Δ for the three rounds, and process reduction would produce a scenario in which the conflict manager should advise the Seller to lower the value of his proposal, in order to become more collaborating in the first round. After the first round actually takes place, the true negotiation style is observed and it triggers a fact arrival phase. The previous scenarios are revised and incorporate the newly arrived information. At the same time, the observed negotiation style is used to condition the Bayesian probabilistic model with a new MAP query, which results in a new default set Δ. This new default set is used in the default revision phase to generate a new set of scenarios and, thus, new recommendations for the conflict manager to give. Supposing the observed negotiation style of the Seller in round one was *collaborating*, as an effect of the recommendation provided by the conflict manager, this would be used to condition parameter *Round*1 in the Seller BN, through the second MAP query given as an example in Section 3.3. This would generate a new default set with the style *collaborating* for round two and *collaborating* or *accommodating* for round three. With default revision, this new default can be used to produce scenarios that let the conflict manager know that it should not intervene since predicted negotiation style is according to the goal it set out for the Seller. This whole process repeats itself throughout the ensuing rounds until the negotiation is brought to a conclusion.

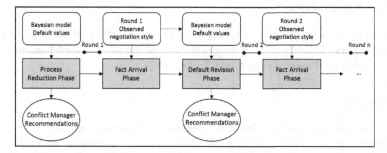

Fig. 2: Workflow carried out by the Conflict Manager based on a Speculative Computation framework.

The goal here is to model a dynamic and flexible conflict manager. For this purpose, BNs are combined with Speculative Computation. While BNs have the task of producing default values, Speculative Computation structures the reasoning process. This allows a quick adaptation to changes in the negotiation style throughout the rounds. This setting also provides a mechanism for the management of conflicts through the anticipation of competing postures.

5   Conclusion

This study set out to determine a way to deal with incomplete information, namely information regarding the conflict style used by parties. Thus, firstly we aim to identify and apply a set of methods for classifying conflict styles with incomplete information; secondly, we aim to validate this approach opposing data collected from a web-based negotiation game. The findings suggest that in general it is now possible to state that our approach copes with incomplete information by producing valid conflict styles default values; and anticipating competing postures through the dynamic generation of recommendations for a conflict manager. So, facing these results, we can claim that this work advances the state of the art in that such an analysis has not previously been undertaken. Meanwhile, due to the small sample size used in the current study, some caution must be taken when interpreting the results of the statistical analysis presented and underpinning the conclusions. Additionally, limitations of the current research must be pointed out. First, the participants were recruited from a particular population- a population that may limit the generalizability of the results. Admittedly, the participants of the experiment may not be representative of conflict parties in general. Consequently, we are unable to demonstrate the causality of the variables conclusively. Finally, returning to the question posed at the beginning of this study, it is now possible to state that is possible to cope with incomplete information to enable a more accurate characterization of conflict styles in a resolution process. However, more research is needed to provide more definitive evidence.

Acknowledgements

This work has been supported by COMPETE: POCI-01-0145-FEDER-007043 and FCT - Fundação para a Ciência e a Tecnologia (Portuguese Foundation for Science and Technology) within the Project Scope UID/CEC/00319/2013.

References

1. Andrew Stranieri and John Zeleznikow. Tools for intelligent decision support system development in the legal domain. In *12th IEEE International Conference on Tools with Artificial Intelligence (ICTAI 2000), 13-15 November 2000, Vancouver, BC, Canada*, pages 186–189, 2000.
2. Davide Carneiro, Marco Gomes, Paulo Novais, and José Neves. chapter Developing Dynamic Conflict Resolution Models Based on the Interpretation of Personal Conflict Styles, pages 44–58. Springer Berlin Heidelberg, Berlin, Heidelberg, 2011.
3. Finn V. Jensen. Bayesian networks. *Wiley Interdisciplinary Reviews: Computational Statistics*, 1(3):307–315, 2009.
4. Marco Gomes, Tiago Oliveira, Davide Carneiro, Paulo Novais, and José Neves. Studying the effects of stress on negotiation behavior. *Cybernetics and Systems*, 45(3):279–291, 2014.
5. Dimitris Margaritis. *Learning Bayesian network model structure from data*. PhD thesis, Carnegie-Mellon University, Pittsburgh, 2003.
6. Laura E. Tsamardinos, Ioannisand Brown and Constantin F. Aliferis. The max-min hill-climbing bayesian network structure learning algorithm. *Machine Learning*, 65(1):31–78, 2006.
7. Marco Scutari. Learning Bayesian networks with the bnlearn R package. *Journal of Statistical Software*, 35(3):1–22, 2010.
8. Kevin Korb and Ann Nicholson. *Bayesian Artificial Intelligence*. CRC Press, London, 2 edition, 2011.
9. Ken Satoh, Philippe Codognet, and Hiroshi Hosobe. Speculative constraint processing in multi-agent systems. In Jaeho Lee and Mike Barley, editors, *Intelligent Agents and Multi-Agent Systems*, volume 2891 of *Lecture Notes in Computer Science*, pages 133–144. Springer Berlin Heidelberg, 2003.
10. T Oliveira, K Satoh, J Neves, and P Novais. Applying Speculative Computation to Guideline-Based Decision Support Systems. In *IEEE 27th International Symposium on Computer-Based Medical Systems 2014 (CBMS)*, pages 42–47, 2014.
11. Robert Castelo and Tomáas Kocka. On inclusion-driven learning of bayesian networks. *J. Mach. Learn. Res.*, 4:527–574, December 2003.
12. Ioannis Tsamardinos, Constantin F Aliferis, Alexander R Statnikov, and Er Statnikov. Algorithms for large scale markov blanket discovery. In *Proceedings of the Sixteenth International Florida Artificial Intelligence Research Society Conference*, volume 2, pages 376–381. AAAI Press, 2003.

Forming Classes in an e-Learning Social Network Scenario

Pasquale De Meo, Lidia Fotia, Fabrizio Messina, Domenico Rosaci, and Giuseppe M. L. Sarné

Abstract Online Social Networks are suitable environments for e-Learning for several reasons. First of all, there are many similarities between social network groups and classrooms. Furthermore, trust relationships taking place within groups can be exploited to give to the users the needed motivations to be engaged in classroom activities. In this paper we exploit information about users' skills, interactions and trust relationships, which are supposed to be available on Online Social Networks, to design a model for managing formation and evolution of e-Learning classes and providing suggestions to a user about the best class to join with and to the class itself about the best students to accept. The proposed approach is validated by a simulation which proves the convergence of the distributed algorithm discussed in this paper.

1 Introduction

E-Learning represents a great opportunities for students to improve their knowledge, as well as to get access to learning material [3]. As users progresses depend on their personal attitudes, background and the learning environment, suitable class formation processes can support e-learning [7], in order to to create comfortable

P.De Meo
DICAM, University of Messina, Italy – e-mail: demeo@unime.it

L.Fotia, D.Rosaci
DIIES, University Mediterranea of Reggio Calabria, Italy – e-mail: \{lidia.fotia,domenico.rosaci\}@unirc.it

G.M.L.Sarné
DICEAM, University Mediterranea of Reggio Calabria, Italy – e-mail: sarne@unirc.it

F.Messina
DMI, University of Catania, Italy – e-mail: messina@dmi.unict.it

© Springer International Publishing AG 2017   173
C. Badica et al. (eds.), *Intelligent Distributed Computing X*,
Studies in Computational Intelligence 678, DOI 10.1007/978-3-319-48829-5_17

environments. Furthermore, dynamics of formation [10], evolution [13] and failure [14] of OSN (Online Social Networks) groups are similar to those of e-Learning classes, so that information related to OSN groups may be used to drive classes formation tasks [16]. Therefore, exploiting existing users' relationships within OSN, can result in an effective number of suggestions for students (resp., classes) about the best classes (resp., students) to join with (resp., to accept into a class) [17] in order to increase the probability that students' expectations are well satisfied. In addition to the common practice of suggesting OSN groups to users by selecting like-minded people [2], studies on social communities confirm that the larger the level of reciprocal trust among the users of a community, the larger their interest to start interactions [4, 5].

Given the premises above, our contribution is represented by a solution to manage the composition of e-Learning classes by using user information – available on OSNs – which are combined in a unique measure, named *convenience*, to suggest the best class to join with or to leave to a user. The first information taken into account is given by the student background on a set of topics of interest. This information is fundamental for the class teaching-homogeneity, and it is based on the number of interactions (in terms of "demand" and "offer") they carried out with other OSN users, e.g. meetings, discussions, etc. In addition, trust relationships are combined (i.e. weighted) with the level of learners' satisfaction and the relevance they assign to them. The model discussed in this work is supported by a multi-agent architecture driving learners' interactions in a fully decentralized manner. A distributed procedure, named Class Formation (CF), allows learner/class software agents to cooperate to form classes by exploiting the convenience measure. To test the proposed approach, a number of experimental trials has been arranged. Experimental results show that the approach is effective in organizing classes and improve the quality of interactions among learners.

The rest of the paper is organized as follows. Section 2 deals with related literature, while section 3 introduces the reference multi-agent scenario, along with the definition of the different metrics which lead to the definition of the convenience measure. Then, section 4 introduces the adopted multi-agent architecture, while section 5 illustrates the two CF algorithms. Experimental results are outlined in section 6. Finally, in section 7 conclusions are drawn.

2 Related Work

As widely discussed in [15], there is a strict relation between e-Learning, i.e. Computer-Supported Collaborative Learning (CSCL) or Technology Enhanced Systems (TES) and OSNs. Therefore, group/class formation is an important task to promote learning activities. A recent work [11] discusses the role of group formation as a first step to design a CSCL on which students can learn and effectively participate to the class activities, and a recent survey on algorithms for group/class formation in CSCL is presented in [6]. From the analysis of about 250 works, the authors conclude that about 50 studies are strictly concerned to group formation in collaborative e-Learning and, in particular, about 20% of them are based on probabilistic models. Some other studies rely on Genetic Algorithms (GA), swarm intelligence, data mining (e.g., k-means), Bayesian networks, and machine learning. In [1] two tools are presented, PQAS and GroupOrganizer, designed to support group formation for collaborative learning in the specific field of computer programming, which have been proved to be effective in increasing students learning in terms of programming style and workgroup skills. Another interesting work deals with strategies for group formation based on individual behaviors [12], on which participation behaviors are analyzed during whole class discussions, with particular emphasis to active participation in small groups. The authors conclude that the instructors would use these information to allocate initial classes into small groups heterogeneously. Another recent survey [8] dealt with the recommender systems for Technology Enhanced Learning (TEL). As we discuss in the next section, in our work we include a recommender system for learners, focusing on the interactions occurring between them with particular emphasis on those involving e-Learning interactions.

3 The Basic OSN scenario and Behavioral, Trust and Convenience Measures

The reference OSN e-Learning scenario is modeled as a directed graph whose nodes represent users and encode their learning characteristics, while each directed edge between a pair of nodes encodes their past interactions. Below we assume that: (i) $\mathcal{N}$ is the OSN members space ($\|\mathcal{N}\| = N$); (ii) C is the set of classes ($\|C\| = C$), each one focused on a topic; (iii) each class $c \in C$ consists of a number of learners, and at least a teacher; (iv) each learner $u_i \in \mathcal{N}$ is assisted by a software agent [9] a_i; (v) Each class manager is assisted by a software agent A_i. Besides, the class composition is based on: (i) *behavioral measure*, referring to the past learner' interactions; (ii) *trust measure*, considering the trust level placed by an OSN member on another OSN member. Such measures are combined in a unique measure denoted as *convenience*.

Behavioral Measures. Assuming that each learner is characterized by a set of *skills* subject to changes and constraints, one of the key concept is that learners can interact with peers of their own classes. Therefore, a behavioral measure describes the concept that learners join with a class also for the capabilities of its members while the class managers are interested to include those users holding attractive skills to offer by means of interactions.

A class c is defined as a tuple $\langle S, W, V_c, o \rangle$ where: $S = \{s_1, \ldots, s_m\}$ is the skill set required by the class manager for c; $W = \{w_1, \ldots, w_n\}$ represent a number of weights used by the class agent to evaluate the students' skills (their sum closes to 1); V_c is the minimum overall skill grade needed to join with c, defined as the estimated grade of an OSN user computed over a specific skill set S. For a user u_k, the overall skill is defined as $V(k,S) = \sum_{i=1}^{m} w_i \cdot g(k,s_i)$, where $g(k,s_i) \in [0,1] \in \mathbb{R}$ is the u_k's knowledge grade coming from the evaluation of the skill $s_i \in S$, o is the reference topic of c.

Let be H the "historical" attitude of the OSN user u_k to require and/or offer interactions for a skill set for a generic class, as follows

$$H(k) = \alpha \cdot H(k)^{(t-1)} + (1-\alpha) \cdot \overline{H}(k)^{(t)} \qquad \overline{H}(k)^{(t)} = \frac{1 - \widetilde{V}(k)_{req}^{(t)}}{1 - \widetilde{V}(k)_{req}^{(t)} + \widetilde{V}(k)_{off}^{(t)}}$$

$$\widetilde{V}(k)_{req}^{(t)} = \frac{1}{N_{req}} \cdot \sum_{i=1}^{N_{req}} V(S_i^{req}, k) \qquad \widetilde{V}(k)_{off}^{(t)} = \frac{1}{N_{off}} \cdot \sum_{i=1}^{N_{off}} V(S_i^{off}, k)$$

where the new value of $H \in [0,1] \in \mathbb{R}$ at the time t combines the values computed at time $t-1$ and t, with $\alpha \in [0,1]$, while $\widetilde{V}(k)_{off}$ is the evaluation of the offered interactions for a number of skills subset $S_i \subset S$, and $1 - \widetilde{V}(k)_{req}$ characterizes the requested interactions. For the generic class c_j, the *Class Behavior* $B(j) \in [0,1]$, can be obtained by averaging over the behavioral measures $H(k)$ of its components:

$$B(j)^{(t)} = \frac{1}{\|c_j\|} \sum_{k=1}^{\|c_j\|} H(k)^{(t)}$$

Trust Measures. This measure is based on the concept of *trust* computed by combining reliability and reputation. Reliability is based on the "quality" of direct knowledge, while the other is an indirect measure perceived by the whole user (agent) community. Due to any interaction (e.g., check homework, asking explanations, etc) between two learners u_p and u_r, we assume that agents a_k and a_r will give feedbacks about the quality of the interactions. Trust $\tau_{p,r}$ is computed by combining reliability (let be $\eta_{p,r}$) and reputation (let be $\rho_{p,r}$) by the coefficient $\beta_{p,r} \in [0,1]$, as

$$\tau_{p,r} = \beta_{p,r} \cdot \eta_{p,r} + (1 - \beta_{p,r}) \cdot \rho_{p,r}$$

Computation of Reliability and Reputation. The reliability $\eta_{p,r} \in [0,1]$ is computed by the agent a_p about the agent a_r:

$$\eta_{p,r}^{(t)} = \vartheta_{p,r} \cdot \sigma_{p,r} + (1 - \vartheta_{p,r}) \cdot \eta_{p,r}^{(t-1)}$$

where $\vartheta_{p,r}$ weights (*i*) the feedback $\sigma_{p,r} \in [0,1] \in \mathbb{R}$ for the last interaction between u_p and u_r at time-step t and (*ii*) the value of $\eta_{p,r}$ computed at time-step $(t-1)$. The reputation measure $\rho_{p,r} \in [0,1]$ is computed by u_p (a_p) with respect to u_r (a_r). More in detail, when u_p is interested to calculate the reputation of u_r (a_r) his/her associated agent a_p automatically asks an opinion about u_r (a_r) to one or more agents a_q associated with OSN users. We assume that the opinion of the generic a_q consists of a function $f(\cdot)$ of its own trust measure about a_r ($\tau_{q,r}$). Finally, the overall reputation $\rho_{p,r}$ is computed as follows:

$$\rho_{p,r} = \frac{1}{l} \sum_{k=1}^{l} f(\tau_{q,r})$$

Convenience Measures. It measures the *convenience* for a user to join with the class c_j. Let $\gamma_{i,j}$ and $\eta_{j,k}$ be the conveniences of the user u_k to join with the class c_j and of the class c_j to accept the request of u_k computed as

$$\gamma_{k,j} = |1 - (H(k) - B(j))| \cdot \frac{1}{\|c_j\|} \sum_{i \in c_j} \tau_{k,i} \qquad \eta_{j,k} = |1 - (H(k) - B(j))| \cdot \frac{1}{\|c_j\|} \sum_{i \in c_j} \tau_{i,k}$$

where $\|c_j\|$ is the number of users (agents) affiliated with c_j. These measures increase with the difference between a_i and c_j behaviors. The asymmetric part is due to the different trust measures $\tau_{k,i}$ and $\tau_{i,k}$, with $a_i \in c_j$. For this reason the procedure described in Section 5 is distributed among the agents assisting learners and class administrators to reach a balance in terms of convenience among all the considered actors of the proposed OSN e-Learning scenario.

4 The Dynamic Management Support of e-Learning Classes

In our approach, OSN users (i.e., learners) are supported by intelligent software agents [18] driven by the measures defined in Section 3.

An Agent Directory Facilitator (ADF) is associated with the whole agent framework and integrated into the whole OSN e-Learning platform, in order to allow agents to send messages by means of a yellow page service.

Learners Profiles. We assume that the generic learner or class agent, respectively associated with a learner or with a class manager, administrates the *profile* of its owner, which consists of the *Working Data* (*WD*) storing the data needed to be operative on the platform and the *Behavioral Data(BD)* concerning the behavioral measures of a student for the learner agent or of the class for the class manager.

The Learner Agent Behavior. The learner agent behavior consists of several tasks, which are described in Section 5. Specifically, let u_k be the generic learner and a_k the associated software agent delegated to execute the following activities: (*i*) *updating behavioral and reliability measures* after each u_k's interaction with his/her classmates; (*ii*) *updating the convenience measure* when change the reliability measure (note: a_k will update also the correspondent reputation measure); (*iii*) *Behavioral and trust measures will be sent* to the agent of the classes, to which the learner has joined with, each time they are recalculated. (*iv*) *assisting u_k in deciding about joining with or leaving classes.*

The Class Agent Behavior. Let c_j be the generic e-Learning class and A_j the associated software agent delegated to assist the class manager to administrate the class itself. The tasks performed by A_j are the following: (i) a learner agent containing updated behavioral and/or convenience measure will trigger agent A_j to *update behavioral and/or convenience measures* for the class itself; (ii) If the behavioral measure of the entire class c_j has changed, A_j will send the updated measure to all the agents the students of the class c_j; (iii) A_j will assist class administrator of the class c_j to decide about the requests coming from software agents to join with or leave classes (see Section 5 for the distributed class formation procedure).

5 The distributed procedure for Class Formation (CF)

This Section reports the design of a distributed procedure, named CF (Class Formation), which allows learner agents and class agents to use the convenience measure defined in Section 3 to manage class composition.

The CF procedure differs for learning and class agents, respectively, as described below. To this aim, let T be the time between 2 consecutive executions (steps) of the CF procedure. We assume also that agents can query a distributed database named CR (Class Repository) linking each formed class with a topic $o \in O$ (e.g. "English"). *The CF procedure performed by the e-learning agents.* Let X_n be the set of the classes which the agent a_n is affiliated to, X_n the current class set to which agent a_n is affiliated. Let N_{MAX} be the maximum number of classes which an e-Learning agent can analyze at time t and ξ_n be a timestamp threshold and $\chi_n \in [0, 1] \in \mathbb{R}$ be a threshold fixed by the agent a_n. We suppose that $N_{MAX} \geq |X_n|$ and that a_n stores into a cache the class profile P_c of each class contacted in the past and the timestamp d of the execution of the CF procedure for that class.

The *CF procedure performed by the learner agent* a_n is represented by the pseudo code listed in the left of Fig. 1. In the procedure a_n tries to improve the convenience value computed on the classes already joined by its user. Initially, the convenience values are refreshed if older than the threshold ξ_n (lines 1-4). Then, candidate classes are sorted in decreasing order basing on their convenience value (line 5) and those having convenience larger than ξ_n are selected. Agent a_n will try to join with this class and could leave other classes. This step attempts to improve the convenience over a target level (represented by ξ_n). As we discuss later, each class agent may reject the join request of an agent or may send a *leave* message to any of its own students if the computed convenience value does not exceed a given threshold.

The CF procedure performed by the class agent is represented by the pseudo code on the right of Fig.1. Let K_c be the set of the agents affiliated to the class c, where $\|K_c\| \leq K_{MAX}$, being K_{MAX} the maximum number of learners allowed to be within the class c, and suppose that the class agent A_c stores into its cache the profile P of each user u managed by his/her learner agent $a \in K_c$, as well as the timestamp d_u of its acquisition. Moreover, let ω_c (a timestamp) and $\pi_c \in [0, 1] \in \mathbb{R}$ be a threshold fixed by A_c. The procedure performed by A_c is activated by any join request sent by a learner agent a_r. Initially, K_{MAX} represents the maximum number of student for any class [1]. If this maximum is reached, no any student is accepted to join with the class, until another student will leave the class. By lines 4-8 the class agent asks the updated profile of the class members, then the convenience $\gamma_{c,a}$ is computed for all these agents (line 12) and a new, sorted set $K_{good} \subset \{K_c \bigcup a_r\}$ is built. Thereafter, the class agent will send a *leave* message to all the learner agents a showing a convenience $\gamma_{c,a}$ lower than π_n.

6 Experiments

In order to prove the effectiveness of the *CF* algorithm in managing classes, we define $\mathbf{AC_j}$ – as measures of the internal convenience measured for a class c_j –, and $\mathbf{MAC}$ and $\mathbf{DAC}$ for the whole set of classes:

$$AC_j = \frac{\sum_{a_i \in c_j} \eta_{j,k}}{\|c_j\|} \qquad MAC = \frac{\sum_{c_j \in C} AC_j}{\|C\|} \qquad DAC = \sqrt{\frac{\sum_{c_j \in C}(AC_j - MAC)^2}{\|C\|}}$$

Input:
$X_n, O_n, N_{MAX}, \xi_n, \chi_n;$
$\quad Y = \{c \in C\}$ a random set of classes : $|Y| \le N_{MAX}$,
$X_n \bigcap Y = \{0\}, \quad Z = (X_n \bigcup Y)$

1: **for** $c \in Z : d_c > \xi_n$ **do**
2:     Send a message to A_c to retrieve the profile P_c.
3:     Compute $\gamma_{u_n,c}$
4: **end for**
5: Let be $L_{good} = \{c_i \in Z : i \le j \rightarrow \gamma_{u_n,c_i} \ge \gamma_{u_n,c_j}\}$,
    with $|L_{good}| = N_{MAX}$
6: $j \rightarrow 0$
7: **for** $c \in L_{good}$ **do**
8:     **if** $(c \notin X_n \wedge \gamma_{u_n,i} > \chi_n)$ **then**
9:         send a join request to A_c
10:         **if** A_c accepts the request **then**
11:             $j \rightarrow j+1$
12:         **end if**
13:     **else**
14:         $j \rightarrow j+1$
15:     **end if**
16: **end for**
17: **for** $c \in \{X_n - L_{good}\}$ **do**
18:     Sends a leave message to c
19:     $j \leftarrow j-1$
20:     **if** $(j == 0)$ **then**
21:         **break**
22:     **end if**
23: **end for**

Input:
$K_c, K_{MAX}, W, \omega_c, \pi_n, a_r, Z = K_c \bigcup \{a_r\};$

1: **if** $(V(r, S_c) < V_c \vee |K_c| \ge K_{MAX})$ **then**
2:     Send a reject message to a_r
3: **else**
4:     **for** $a \in K_c$ **do**
5:         **if** $d_u \ge \omega_c$ **then**
6:             ask to a its updated profile
7:         **end if**
8:     **end for**
9:     **for** $a \in Z$ **do**
10:         compute $\eta_{c,a}$
11:     **end for**
12:     Let be $K_{good} = \{a \in Z : \gamma_{c,a} \ge \pi_c\}$
13:     **for** $a \in K_c - K_{good}$ **do**
14:         send a *leave* message to a.
15:     **end for**
16:     **if** $a_r \in K_{good}$ **then**
17:         the request of a_r is accepted
18:     **end if**
19: **end if**

Fig. 1 CF procedure. Left: Learner side, Right: class side

---

[1] We assume for convenience that it is the same for all the classes and topics

The experiments involved three scenarios consisting of 50, 100, and 200 e-Learning classes, as summarized in Table 1.

Sc.	$|C|$	$|U|$	K_{Max}	N_{Max}
1	50	200		
2	100	400	20	5
3	200	800		

τ_t	τ_{ut}	V_{Req}, V_{off}
$v = 0.8, \sigma = 0.3$	$v = 0.3, \sigma = 0.2$	$v = 0.5, \sigma = 0.2$

Table 1 Simulation parameters

Users profiles were constructed by assuming that the 20% of the OSN members have an "untrusted" behavior, while the remaining 80% assume a behavior that can be considered "trusted", while behavioral coefficients H (see Section 3), have been computed around some average values of V_{req} and V_{off}. All the values above, i.e. trust (τ) and behavioral coefficient V_{req} and V_{off}, have been generated by sampling from a normal distribution with average v and standard deviation σ, as specified in table 1. While τ_t represents the generated values for "trusted" users (i.e. that users that show, in average a trusted behavior), τ_{ut} represents trust values for "untrusted". For this first set of experiments, we set the ratio $r = \dfrac{K_{max} \cdot |C|}{N_{max} \cdot |U|} = 1$. Furthermore, the starting composition of classes was set as random.

Sc.	T	MAC	DAC
1	0	0.63	0.12
	20	0.67	0.10

Sc.	T	MAC	DAC
2	0	0.62	0.12
	20	0.66	0.10

Sc.	T	MAC	DAC
3	0	0.62	0.12
	20	0.67	0.12

Table 2 Results with $r = 1.0$

Table 2 shows the results obtained for the three scenarios reported in Table 1, with the initial value of MAC/DAC (epoch T=0) and the final one (epoch T=20). In this case we observe that the improvement, in terms of MAC, at the end of the experiments, is in the order of 8% for all the tested configuration. We extended the test above in order to verify that, when the ratio $\dfrac{K_{max} \cdot |C|}{N_{max} \cdot |U|}$ does not change, and there are no relevant variation in the improvement the MAC. Therefore, we performed a further set of experiments, which are described below, by means of a relevant variation of the ratio r.

In particular, for this second set of experiments, ratio r ranges from 0.1 to 0.9, as shown in table 3. A value $r < 1$ indicates that users, in overall, can join more places ($N_{max} \cdot |U|$) than the total available in average ($K_{max} \cdot |C|$). By looking at Table 3, it can be observed that the best improvement, in terms of *MAC*, is obtained for a value of r between 0.4 and 0.6. This result can be explained as follows. On one hand, finding a class to improve the personal convenience γ is a bit more difficult for the single user when $r < 1$, therefore the CF algorithm is effective in improving the MAC with respect to a random composition of classes. Nevertheless, when $r << 1$ the

r	T MAC	DAC	r	T MAC	DAC	r	T MAC	DAC
0.1	0 0.61	0.07	0.4	0 0.60	0.04	0.7	0 0.63	0.09
	20 0.61 (+0%)	0.08		20 0.70 (+20%)	0.1		20 0.69 (+9%)	0.08
0.2	0 0.59	0.02	0.5	0 0.60	0.08	0.8	0 0.63	0.09
	20 0.63 (+8%)	0.08		20 0.73 (+21%)	0.06		20 0.68 (+8%)	0.08
0.3	0 0.60	0.03	0.6	0 0.60	0.07	0.9	0 0.62	0.11
	20 0.69 (+14%)	0.04		20 0.7 (+16%)	0.09		20 0.67 (+8%)	0.10

Table 3 Results with $r = 0.1$ to $r = 0.9$

improvements, in terms of *MAC* is lower than in the case 0.4−0.6, and is comparable to the improvement given when the values of r is closed to 1. In overall, these results clearly point out that the CF algorithm introduces a significant increment of the convenience of the classes, especially when organizing classes for e-learning is difficult due to limited "seats".

7 Conclusions and future work

Due to many similarities between social network groups and classrooms, in this work we exploit data related to skills and interactions between e-learners and trust relationships which are established within groups in ONSs.

In particular, we presented a model to manage formation and evolution of e-Learning classes in OSNs context. In particular, a distributed algorithm aimed at providing suggestions to a user about the best class to join with and to the class itself about the best students to accept has been designed. The proposed approach has been validated by a number of simulations which prove the convergence of the distributed algorithm for class formations.

As a future work, we aim at providing a further experimental analysis aimed at moving a step forward in this research. First of all, starting from the fact that the distributed algorithm allows class administrators to improve the average convenience of the classrooms, we aim at prove that the obtained class compositions actually hold a better "quality of interactions" among e-learners. Moreover, since the trust evaluation system is involved in the class formation, it should be improved with a number of countermeasures by which malicious behaviors aimed at improving the mutual convenience — in order to join certain classes — are neutralized.

Acknowledgment

This work has been partially supported by the Program "Programma Operativo Nazionale Ricerca e Competitività" 2007-2013, Distretto Tecnologico CyberSecurity funded by the Italian Ministry of Education, University and Research.

References

1. Juan Manuel Adán-Coello, Carlos Miguel Tobar, Eustáquio São José de Faria, Wiris Serafim de Menezes, and Ricardo Luís de Freitas. Forming groups for collaborative learning of introductory computer programming based on students programming skills and learning styles. *International Journal of Information and Communication Technology Education (IJICTE)*, 7(4):34–46, 2011.
2. V. Agarwal and K.K. Bharadwaj. A collaborative filtering framework for friends recommendation in social networks based on interaction intensity and adaptive user similarity. *Social Network Analysis and Mining*, 3(3):359–379, 2013.
3. Ruth C Clark and Richard E Mayer. *E-learning and the science of instruction: Proven guidelines for consumers and designers of multimedia learning*. John Wiley & Sons, 2011.
4. A. Comi, L. Fotia, F. Messina, G. Pappalardo, D. Rosaci, and G. M. L. Sarné. Forming homogeneous classes for e-learning in a social network scenario. In *Intelligent Distributed Computing IX*, pages 131–141. Springer, 2016.
5. Antonello Comi, Lidia Fotia, Fabrizio Messina, Domenico Rosaci, and Giuseppe ML Sarnè. A qos-aware, trust-based aggregation model for grid federations. In *On the Move to Meaningful Internet Systems: OTM 2014 Conferences*, pages 277–294. Springer, 2014.
6. Wilmax Marreiro Cruz and Seiji Isotani. Group formation algorithms in collaborative learning contexts: A systematic mapping of the literature. In *Collaboration and Technology*, pages 199–214. Springer, 2014.
7. M.I. Dascalu, C.N. Bodea, M. Lytras, P.O. De Pablos, and A. Burlacu. Improving e-learning communities through optimal composition of multidisciplinary learning groups. *Computers in Human Behavior*, 30:362–371, 2014.
8. M. Erdt, A. Fernandez, and C. Rensing. Evaluating recommender systems for technology enhanced learning: A quantitative survey. *Learning Technologies, IEEE Transactions on*, 8(4):326–344, Oct 2015.
9. S. Franklin and A. Graesser. Is it an agent, or just a program?: A taxonomy for autonomous agents. In *Intelligent agents III Agent Theories, Architectures, and Languages*, pages 21–35. Springer, 1997.
10. P. Grabowicz, L. Aiello, V. Eguiluz, and A. Jaimes. Distinguishing topical and social groups based on common identity and bond theory. In *Proc. of the ACM Int. Conf. on Web Search and Data Mining*, pages 627–636. ACM, 2013.
11. Seiji Isotani, Akiko Inaba, Mitsuru Ikeda, and Riichiro Mizoguchi. An ontology engineering approach to the realization of theory-driven group formation. *International Journal of Computer-Supported Collaborative Learning*, 4(4):445–478, 2009.
12. Namsook Jahng and Mark Bullen. Exploring group forming strategies by examining participation behaviours during whole class discussions. *European Journal of Open, Distance and E-Learning*, 2012.
13. S.R. Kairam, D.J. Wang, and J. Leskovec. The life and death of online groups: Predicting group growth and longevity. In *Proc. of the 5th ACM int. conf. on Web search and data mining*, pages 673–682. ACM, 2012.
14. R.E. Kraut and A.T. Fiore. The role of founders in building online groups. In *Proc. of the 17th ACM conf. on Computer Supported Cooperative Work & Social Computing*, pages 722–732. ACM Press, 2014.
15. Frank Rennie and Tara M Morrison. *E-learning and social networking handbook: Resources for higher education*. Routledge, 2013.
16. G. Sakarkar, S.P. Deshpande, and V.M. Thakare. An online social networking architecture using context data for effective e-learning systems. In *Emerging Research in Computing, Information, Communication and Applications (Vol 1)*, pages 33–39, 2014.
17. V. Vasuki, N. Natarajan, Z. Lu, B. Savas, and I. Dhillon. Scalable affiliation recommendation using auxiliary networks. *ACM Trans on Intelligent Systems and Technology*, 3(1):3, 2011.
18. Michael Wooldridge and Nicholas R Jennings. Intelligent agents: Theory and practice. *The knowledge engineering review*, 10(02):115–152, 1995.

Part VI

Optimization

Scheduling Optimization in Grid with VO Stakeholders' Preferences

Victor Toporkov, Anna Toporkova, Dmitry Yemelyanov,
Alexander Bobchenkov, and Alexey Tselishchev

Abstract The problem of intelligent Grid computing and job-flow scheduling with regard to preferences given by various groups of virtual organization (VO) stakeholders (such as users, resource owners and administrators) is studied. A specific flexible resources share algorithm is proposed for job-flow scheduling which enables to achive a balance between the VO stakeholders' conflicting preferences and policies. This approach provides greater VO scheduling fairness, improves the overall quality of service and resource load efficiency. Two different metrics are introduced to find a scheduling solution balanced between VO stakeholders. Experimental results prove that the cyclic scheduling scheme allows establishing efficient cooperation between different VO stakeholders even if their goals and preferences are contradictory.

Key words: intelligent Grid, stakeholders, preferences, scheduling, virtual organization, economic models.

Victor Toporkov · Dmitry Yemelyanov · Alexander Bobchenkov
National Research University "MPEI"
ul. Krasnokazarmennaya, 14, Moscow, 111250, Russia
{ToporkovVV,YemelyanovDM,BobchenkovAV}@mpei.ru

Anna Toporkova
National Research University Higher School of Economics
ul. Myasnitskaya, 20, Moscow, 101000, Russia
AToporkova@hse.ru

Alexey Tselishchev
European Organization for Nuclear Research (CERN)
Geneva, 23, 1211, Switzerland
Alexey.Tselishchev@cern.ch

© Springer International Publishing AG 2017   185
C. Badica et al. (eds.), *Intelligent Distributed Computing X*,
Studies in Computational Intelligence 678, DOI 10.1007/978-3-319-48829-5_18

1 Introduction and Related Works

In distributed computing with a lot of different participants and contradicting requirements the well-known efficient approaches are based on economic principles [2, 3, 4, 5, 6, 7, 8]. Two established trends may be outlined among diverse approaches to distributed computing. The first one is based on the available resources utilization and application level scheduling [2]. As a rule, this approach does not imply any global resource sharing or allocation policy. Another trend is related to the formation of user's virtual organizations (VO) and job flow scheduling [12]. In this case a metascheduler is an intermediate chain between the users and local resource management and job batch processing systems.

Uniform rules of resource sharing and consumption, in particular based on economic models, make it possible to improve the job-flow level scheduling and resource distribution efficiency.

In most cases, VO stakeholders pursue contradictory goals working on Grid. VO policies usually contain mechanisms of interaction between VO stakeholders, rules of resource management and define user shares [3, 9]. Besides, VO policy may offer optimized scheduling to satisfy both users' and VO common preferences, which can be formulated as follows: to optimize users' criteria or utility function for selected jobs [5, 14], to keep resource overall load balance [10, 22], to have job run in strict order or maintain job priorities [13], to optimize overall scheduling performance by some custom criteria [1, 16, 22], etc.

VO formation and performance largely depends on mutually beneficial collaboration between all the related stakeholders. However, users' preferences and VO common preferences (owners' and administrators' combined) may conflict with each other. Users are likely to be interested in the fastest possible running time for their jobs with least possible costs whereas VO preferences are usually directed to available resources load balancing or node owners' profit boosting. Thus, VO policies in general should respect all members and the most important aspect of rules suggested by VO is their fairness. A number of works understand fairness as it is defined in the theory of cooperative games, such as fair quotas [3, 11], fair user jobs prioritization [13], non-monetary distribution [15].

In many studies VO stakeholders' preferences are usually ensured only partially: either owners are competing for jobs optimizing only users' criteria [4, 5], or the main purpose is the efficient resources utilization not considering users' preferences. Sometimes multiagent economic models are established [2]. They aren't allow to optimize the whole job flow processing.

We consider a job flow scheduling model with a metascheduler as an intermediate chain between the users and a local resource management system [21]. The cyclic scheduling scheme (CSS) [18, 19] has fair resource share in a sense that every VO stakeholder has mechanisms to influence scheduling results providing own preferences. In [18], the combination of CSS and several heuristics for preference-based scheduling was proposed and studied.

A main contribution of this paper is a flexible job-flow scheduling approach that provides a balance between VO common preferences and users' preferences within a CSS framework. The rest of the paper is organized as follows. Section 2 presents a general cyclic scheduling concept. The proposed VO preference based scheduling technique is presented in Section 3. Section 4 contains settings for the experiment and Section 5 contains the simulation results for the proposed scheduling approach. Finally, Section 6 summarizes the paper.

2 Cyclic Scheduling Scheme and the Fair Resource Sharing Concept

Scheduling of a job flow using CSS is performed in time cycles known as scheduling intervals, by job batches [18, 19]. The actual scheduling procedure consists of two main steps. The first step involves a search for alternative scenarios of each job execution or simply alternatives. During the second step the dynamic programming methods [19] are used to choose an optimal alternatives' combination with respect to the given VO criteria. This combination represents the final schedule based on current data on resources load and possible alternative executions.

Alternatives have time (T) and cost (C) properties. In particular, alternatives' time properties include execution CPU time (overall running time), execution start and finish times, total execution runtime. Thus, a common optimization problem may be stated as either minimization or maximization of one of the properties, having other fixed or limited, or involve Pareto-optimal strategy search involving both kinds of properties [6, 12]. For example, total job batch execution time minimization with a restriction on total execution cost $(T \rightarrow \min, \lim C)$.

Alternative executions for each job may be obtained for example as auction-based offers from resource owners [2, 4], as offers from local scheduling systems [5] or by some directed search procedures in a distributed environment [2, 17]. The concept of *fair resource sharing* in VO suggests that all stakeholders of a VO should be able to influence scheduling results in accordance with their own preferences. In order to implement such a mechanism, the resource request alongside with a user job has a custom parameter: a user scheduling criterion. An example may be a minimization of overall running time, a minimization of overall running cost etc. [17]. This parameter describes user's preferences for that specific job execution and expresses a type of an additional optimization to perform when searching for alternatives.

3 Alternative Optimization Based on Users' Preferences

The approach described above has a distinct imbalance. VO administrators in general have much stronger impact on final scheduling results as their criteria are employed at the second step of CSS. In order to recover the balance between VO stakeholders a new type of property is introduced for an alternative: U (user utility)

We consider the following relative approach to represent a user utility function. A job alternative with the minimum user-defined criterion value Z_{min} corresponds to the left interval boundary ($U = 0\%$) of all possible job scheduling outcomes. An alternative with the worst possible criterion value Z_{max} corresponds to the right interval boundary ($U = 100\%$). In the general case for each alternative with value Z of optimization criterion, U is set depending on its position in $[Z_{min}; Z_{max}]$ using the following formula:

$$U = \frac{Z - Z_{min}}{Z_{max} - Z_{min}} * 100\%.$$

Thus, each alternative gets its utility in relation to the "best" and the "worst" optimization criterion values user could expect according to the job's priority. And the more some alternative corresponds to user's preferences the smaller is the value of U. So, then the second step optimization problem could be in form of: $(C \rightarrow \max, \lim U)$, $(U \rightarrow \min, \lim T)$ and so on. Examples of user utility functions for a job with four alternatives and a cost minimization criterion are presented in Table 1.

Table 1 User utility examples for a job with execution cost minimization

Job Execution Alternatives	Execution Cost	Utility
First Alternative	5	0%
Second Alternative	7	20%
Third Alternative	11	60%
Fourth Alternative	15	100%

The formal statement of the second step optimization problem for n batch jobs could be written as:

$$f(\bar{s}) = \sum_{i=1}^{n} f_i(s_j) \rightarrow \text{extr}, u_i(s_j) \leq u_i \leq u^*, \tag{1}$$

where $f_i(s_j)$ is efficiency [21] of an alternative s_j for job i based on VO preferences, $u_i(s_j)$ is a utility for this alternative from the perspective of the user, u_i is a partial user utility sum value (for example for jobs $i, i + 1, , n$ or $i, i - 1, , 1$), u^* is a general limit on user utility for the whole batch, and $\bar{s} = s_1, ..., s_j, ..., s_n$.

Average utility limit $U_a = u^*/n$ which is correlated to the restriction u^* in (1) and can be used to simplify the further analysis. Using this approach one can describe the optimization task for the second step in CSS as follows: minimize the total job batch execution time while on average ensuring the usage of alternatives with 0%-20% deviation from the best possible scheduling outcome ($T \to$ min, lim $U_a = 20\%$).

Furthermore, this approach allows to compare and to balance user versus VO administrator preferences. Indeed, if the "worst" VO optimization criterion value Y_{min} for a job batch is taken with a $U_a \leq 0\%$ average user utility limit (i.e. when only the best user alternatives can be scheduled for execution) and the "best" Y_{max} value with a $U_a \leq 100\%$ limit (when any alternatives can be scheduled for execution), then any value Y, that is coming from the VO optimization problem solution can be expressed in relation to its position on $[Y_{min}; Y_{max}]$. This relative value A will express the degree in which VO administrator interests are met:

$$A = \frac{Y - Y_{min}}{Y_{max} - Y_{min}} * 100\%.$$

Based on U_a and A properties of a job batch scheduling outcome one can tune and balance them for fair resource distribution by setting the appropriate scheduling restriction on U_a.

4 Experiment Settings

An experiment was prepared as follows using a custom distributed environment simulator [20] comprising both application and job-flow scheduling levels.

Virtual organization and computing environment properties are the following. The resource pool includes 100 heterogeneous computational nodes. A specific cost of a node is an exponential function of its performance value (base cost) with an added variable margin distributed normally as $+-0.6$ of a base cost. The scheduling interval length is 600 time quanta. Initially the resources are pre-utilized by 5% to 10% with owner jobs distributed hypergeometrically.

The job batch properties are specified as follows. Jobs number in a batch is 40. Nodes quantity needed for a job is an integer number distributed evenly on $[2; 6]$. Node reservation time is an integer number distributed evenly on $[100; 500]$. Job budget varies in the way that some of jobs can pay as much as 160% of base cost whereas some may require a discount. Every request contains a specification of a custom criterion which is one of the following: job execution runtime, finish time and overall execution cost.

During an experiment a VO and a job batch is generated. Then CSS is applied to solve the optimization problems with a total utility limited. The results can be examined to analyze processes and the efficiency of the fair resource sharing model. These computing environment properties are mainly affect the CSS alternatives search step. However, to study the proposed fair scheduling scheme alternative executions for each job may be obtained in other ways, for example, as auction-based offers [4]. The main difference of the proposed approach is a job batch scheduling implementing both VO and user-based criteria optimization.

The important feature of the present study is how users' preferences comply with a VO common policy and optimization. An experiment is conducted for studies of the following combinations of VO member's preferences:

1. The mixed combination: only a half of jobs comply with VO preferences.
2. The conflict combination: all jobs have custom scheduling criteria that fully contradict VO preferences.

5 Experimental Results

5.1 Utility Function Characteristics

Two series of experiments were carried out for combinations listed above. The conflict combination had the overall job batch execution cost maximized (owners' criterion) with alternatives utility limited ($C \to$ max, lim U). At the same time for all VO users a criterion is set to minimize each job's execution cost. Thus, VO owners' preferences are clearly opposing users' ones.

The mixed combination had the overall cost maximized as well, but only half of jobs voted for cost minimization, the other half had time optimizations (finish time and running time) preferred.

Let us review scheduling results: Fig. 1 shows the average job execution cost for mixed and conflicted combinations together. As it can be observed from the diagram, when there is no constraint on the utility ($U_a = 100\%$), resulting job cost has the maximum possible value, and in case when the utility restriction is the most stringent ($U_a = 0\%$), the resulting cost is the minimum possible. So, by setting the restriction on U_a it is possible to establish some fair scheduling VO policy balanced between common (VO's) and local (users') goals. The result achieved with a $U_a = 100\%$ constraint corresponds to the cycle scheduling scheme described in [21].

A horizontal dashed line in Fig. 1 marks the average between the maximum (VO administrators' preferences) and minimum (VO users' preferences) cost values achieved in the experiment and represents the *compromise* value. This mutual trade-off is achieved in a conflicting combination when limiting U_a to $U_a = 25\%$.

At the same time, cost diagram for a *mixed* configuration is flatter, the minimum value is greater, and the higher values of the optimization criterion

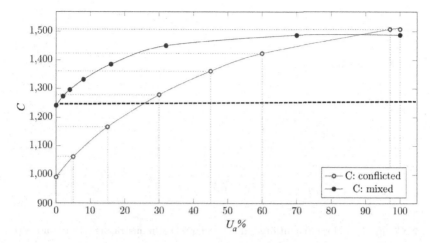

Fig. 1 Average job execution cost comparison in a total job batch cost maximization problem with both *conflict* and *mixed* preferences combinations

(C) are achieved in comparison to a fully conflict preferences combination given the same value of the user U_a constraint. These features allow VO administrators to have more optimization possibilities owing to the tasks with correlating criteria.

5.2 Balancing interests for VO participants

Of particular interest is the mutual compromise problem for VO stakeholders' preferences. As it was shown in Fig. 1 for a conflicting preferences combination, the balance can be achieved when limiting U_a to $U_a = 25\%$, providing value $A = 50\%$. For a general case, it is possible to express VO participants' interests in common, comparable parameters and use relative utility parameters for user alternatives(U_a) and VO job batch scheduling outcomes (A).

The diagram in Fig. 2 shows the dependencies for the scheduling efficiency parameter A in relation to an average user utility U_a. The diagram shows graphs for $C \rightarrow$ max, lim U_a problem in conflicting and mixed preferences combinations and for $T \rightarrow$ min, lim U_a problem in a mixed preferences combination.*Dashed line graph* $A(U_a) = U_a$ *represents user utility function.* Intersection points between graphs, representing user utility and VO scheduling efficiencies, present *a scheduling problem solution*, when both preferences are equally satisfied. For example, if one takes the intersection point of a user utility line with the time execution minimization graph in a mixed configuration (*T:mixed* in Fig. 2), the corresponding solution will ensure 16% scheduling outcome deviation from the best possible criterion value for both users and VO administrators. In a fully conflicting task the similar solution

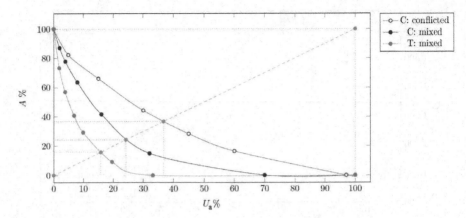

Fig. 2 Comparison of relative utility indicators for VO administrators (A) and users (U_a)

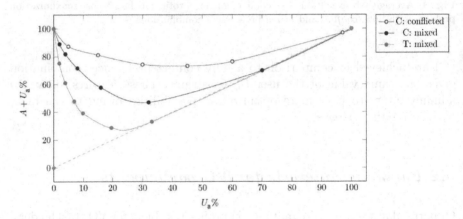

Fig. 3 Sum of relative utility indicators for VO administrators (A) and users (U_a)

is achieved in the point (37%,37%) with the corresponding deviation of 37% from the best possible criteria value. When interests are fully aligned, the solution will be achieved in the point (0%,0%).

It is possible to use other metrics for preferences comparison and balancing. For example, one can use a sum of users' and VO administrators' efficiency parameters as a balancing *sum metric*: $A + U_a$. Fig. 3 contain graphs for the same problems presented in Fig. 2 with a *sum metric*. Minimum points on these graphs correspond to a minimal aggregate deviation from the best possible scheduling outcomes for all the VO participants. With *sum metric* the minimum for the cost maximization in conflicting configuration is achieved with $U_a = 45\%$ and $A = 29\%$. In this case the solution is a bit more favourable for VO administrators compared to a simple conflict solution (point (25%,50%)) and equality solution (37%,37%). It is possible to use other metrics to choose a trade-off solution for the second step in the cyclic scheduling scheme.

6 Conclusions and Future Work

In the present paper, a problem of finding a balance between VO stakeholders' preferences to provide fair resource sharing and distribution is studied. In the framework of the cyclic scheduling scheme a specific flexible resources share algorithm is proposed for job-flow scheduling which enables to achieve a balance between the VO stakeholders' conflicting preferences and policies.

An experimental study included simulation of a job flow scheduling in VO. Different combinations of VO stakeholders' preferences were studied: for example, when some users are in compliance with VO preferences and others being not. Three different compromise solution types which ensures fair resource sharing for scheduling problem are reviewed separately. The experimental results prove the mentioned scheduling scheme suggests tools to establish efficient cooperation between different VO stakeholders even if their goals and preferences are contradictory.

Further research will be related to additional mechanisms development in order to find a scheduling solution balanced between all VO stakeholders.

Acknowledgements This work was partially supported by the Council on Grants of the President of the Russian Federation for State Support of Young Scientists and Leading Scientific Schools (grants YPhD-4148.2015.9 and SS-6577.2016.9), RFBR (grants 15-07-02259 and 15-07-03401), the Ministry on Education and Science of the Russian Federation, task no. 2014/123 (project no. 2268), and by the Russian Science Foundation (project no. 15-11-10010).

References

1. Blanco, H., Guirado, F., Lrida, J.L., Albornoz, V.M.: MIP model scheduling for multi-clusters. In Euro-Par 2012, pages 196-206, Heidelberg, 2012. Springer.
2. Buyya, R., Abramson, D., Giddy., J.: Economic models for resource management and scheduling in Grid computing. J. Concurrency and Computation, 14(5):1507-1542, 2002.
3. Carroll, T., Grosu, D.: Divisible load scheduling: An approach using coalitional games. In Proceedings of the Sixth International Symposium on Parallel and Distributed Computing, ISPDC 07, page 36, 2007.
4. Dalheimer, M., Pfreundt, F., Merz, P.: Agent-based Grid scheduling with Calana. In Parallel Processing and Applied Mathematics, 6th International Conference, pages 741-750. Springer, 2006.
5. Ernemann, C., Hamscher, V., Yahyapour, R.: Economic scheduling in Grid computing. In D. Feitelson, L. Rudolph, and U. Schwiegelshohn, editors, JSSPP, volume 18, pages 128-152. Springer, Heidelberg, 2002.

6. Farahabady, M.H., Lee, Y.C., Zomaya, A.Y.: Pareto-optimal cloud bursting. In IEEE Transactions on Parallel and Distributed Systems, volume 25, pages 2670-2682, 2014.
7. Garg, S., Yeo, C., Anandasivam, C., Buyya, R.: Environment-conscious scheduling of HPC applications on distributed cloud-oriented data centers. J. Parallel and Distributed Computing, 71(6):732-749, 2011.
8. Garg, S.K., Konugurthi, P., Buyya., R.: A linear programming-driven genetic algorithm for meta-scheduling on utility Grids. J. Par., Emergent and Distr. Systems, (26):493-517, 2011.
9. Gulati, A., Ahmad, I., Waldspurger, C.: PARDA: Proportional allocation of resources for distributed storage access. In FAST '09 Proccedings of the 7th conference on File and storage technologies, pages 85-98, California, USA, 2009.
10. Inoie, A., Kameda, H., Touati, C.: Pareto set, fairness, and Nash equilibrium: A case study on load balancing. In Proceedings of the 11th International Symposium on Dynamic Games and Applications, pages 386-393, Arizona, USA, 2004.
11. Kim, K., Buyya, K.: Fair resource sharing in hierarchical virtual organizations for global Grids. In Proceedings of the 8th IEEE/ACM International Conference on Grid Computing, pages 50-57, Austin, USA, 2007. IEEE Computer Society.
12. Kurowski, K., Nabrzyski, K., Oleksiak, A., Weglarz, J.: Multicriteria aspects of Grid resource management. In J. Nabrzyski, Schopf J.M., and J. Weglarz, editors, Grid resource management. State of the Art and Future Trends, pages 271-293. Kluwer Acad. Publ., 2003.
13. Mutz, A., Wolski, R., Brevik, J.: Eliciting honest value information in a batch-queue environment. In 8th IEEE/ACM International Conference on Grid Computing, pages 291-297, New York, USA, 2007. ACM.
14. Rzadca, K., Trystram, D., Wierzbicki, A.: Fair game-theoretic resource management in dedicated Grids. In IEEE International Symposium on Cluster Computing and the Grid (CCGRID 2007), pages 343-350, Rio De Janeiro, Brazil, 2007. IEEE Computer Society.
15. Skowron, P., Rzadca, K.: Non-monetary fair scheduling cooperative game theory approach. In Proceeding of SPAA '13 Proceedings of the twenty-fifth annual ACM symposium on Parallelism in algorithms and architectures, pages 288-297, New York, USA, 2013. ACM.
16. Takefusa, A., Nakada, H., Kudoh, T., Tanaka, Y.: An advance reservation-based co-allocation algorithm for distributed computers and network bandwidth on QoS-guaranteed Grids. In Schwiegelshohn U. Frachtenberg E., editor, JSSPP 2010, volume 6253, pages 16-34. Springer, Heidelberg, 2010.
17. Toporkov, V., Toporkova, A., Tselishchev, A., Yemelyanov, D.: Slot selection algorithms in distributed computing. Journal of Supercomputing, 69(1):53-60, 2014.
18. Toporkov, V., Toporkova, A., Tselishchev, A., Yemelyanov, D., Potekhin, P.: Core heuristics for preference-based scheduling in virtual organizations of utility grids. In Studies in Computational Intelligence, volume 570, pages 321-330. Springer International Publishing, 2015.
19. Toporkov, V., Toporkova, A., Tselishchev, A., Yemelyanov, D., Potekhin, P.:Metascheduling and heuristic co-allocation strategies in distributed computing. Computing and Informatics, 34(1):45-76, 2015.
20. Toporkov, V., Tselishchev, A., Yemelyanov, D., Bobchenkov, A.: Composite scheduling strategies in distributed computing with non-dedicated resources. Procedia Computer Science, 9:176-185, 2012.
21. Toporkov, V., Tselishchev, A., Yemelyanov, D., Potekhin, P.: Metascheduling strategies in distributed computing with non-dedicated resources. In W. Zamojski and J. Sugier, editors, Dependability Problems of Complex Information Systems, Advances in Intelligent Systems and Computing, volume 307, pages 129-148. Springer, 2015.
22. Vasile, M., Pop, F., Tutueanu, R., Cristea, V., Kolodziej,J.: Resource-aware hybrid scheduling algorithm in heterogeneous distributed computing. Future Generation Computer Systems, 51:61-71, 2015.

On the Application of Bio-inspired Heuristics for Network Routing with Multiple QoS Constraints

Miren Nekane Bilbao[1], Cristina Perfecto[1], Javier Del Ser[1,2], and Xabier Landa[1]

[1] University of the Basque Country UPV/EHU, 48013 Bilbao, Spain,
{nekane.bilbao,cristina.perfecto,javier.delser}@ehu.eus
xlanda004@ikasle.ehu.eus
[2] TECNALIA. OPTIMA Unit, E-48160 Derio, Spain,
javier.delser@tecnalia.com

Abstract Since the advent of Telecommunication networks in the early 60's, routing has become a recurrent problem with evergrowing complexity due to the simultaneous share of resources, stringent Quality of Service (QoS) constraints and unmanageable network scales (size, speed and exchanged data volume) by conventional route finding schemes. This paper considers a particular class of routing problems where the route to be found needs to simultaneously fulfill different requirements in terms of e.g. maximum latency, loss rate or any other cost measure. The manuscript delves into the application of the Coral Reefs Optimization and the Firefly Algorithm, two of the latest bio-inspired meta-heuristic techniques reported to outperform other approximative solvers in a wide range of optimization scenarios. Results obtained from Monte Carlo simulations over synthetic network instances will shed light on the comparative performance of these two algorithms, with emphasis on their convergence speed and statistical significance.

Keywords: Constrained Network Routing; Bio-inspired Optimization; Coral Reefs Optimization; Firefly Algorithm

1 Introduction

Nowadays Internet can be conceived as a global mixture of interconnected yet non-connection-oriented network infrastructures, which unreliably process data flows (*best effort*) in the form of packets reaching their destinations through different paths or routes [1]. Since the beginning of its operation as an information exchange network characterized by a very low bandwidth for delay-tolerant applications (e.g. file transfer or e-mail), the Internet has experienced a significant increase in the heterogeneity of traffic carried through its links and constituent nodes, ranging from videoconferencing to video-on-demand or VoIP. The lack of efficient mechanisms for resource reservation in these networks, and the very different quality of service requirements of these multimedia applications with respect to the first deployed Internet services, motivate the growing momentum around mechanisms or techniques that ensure, at different hierarchy levels of the

© Springer International Publishing AG 2017
C. Badica et al. (eds.), *Intelligent Distributed Computing X*,
Studies in Computational Intelligence 678, DOI 10.1007/978-3-319-48829-5_19

OSI model or any other digital communication process, the Quality of Service (QoS) of a given flow of information. In the remainder QoS will refer to the set of performance measures of a particular communication process that quantifies the degree of satisfaction of the user of the service [2].

In this manuscript we will specifically focus on ensuring quality of service at the network level. According to the OSI layered framework, the so-called network layer provides logical mechanisms that allow two remote systems possibly established on different networks to communicate to each other over a certain route. A first classification of routing algorithms can be made by addressing their adaptability to dynamic network environments. On one hand, non-adaptive algorithms make their routing decisions disregarding measurements, estimates of the statistical characteristics of the processed traffic (e.g. arrival rate, burst-like behavior) or the network topology itself. The route to take from one node to another is calculated and programmèd statically before transmission. There are several examples of static routing algorithms:

- Shortest path routing: the communication network is conceived as a graph in which each vertex represents a node of the network, and each edge of the graph a communication link. In order to select the path between a given pair of nodes, the routing algorithm in the graph simply selects the shortest path between them based on a measure of the quality of the link (e.g. the number of hops, the physical distance between them, the estimated transmission delay, the available bandwidth or the cost of communication over that link), for which diverse methods are available (such as the Dijsktra algorithm).
- Flooding: at each node, each incoming packet is sent by each of the output links, except that for which the packet arrived. Flooding generates large quantities of replica or duplicate packets; in fact, in networks with cycles can give rise to a asymptotically infinite amount, unless countermeasures to control this procedure are taken (e.g. by using a hop count taken by each packet). Flooding is not practical in most applications, but is used in those scenarios where communication robustness is a must.

The disadvantage of this family of static routing algorithms is their lack of responsiveness against changing situations, such as network saturation, connection intermittency or network congestion. In dynamic complex networks a moderate degree of cooperation between the nodes should be needed so as to ensure end-to-end resilience over the communication path. This observation motivates the emergence of adaptive routing algorithms, capable of dynamically constructing routes by virtue of the exchange of information between nodes. For example, dynamic distance vector algorithms (used by RIP, as well as in Cisco's IGRP and EIGRP routing protocols [3]) operate by maintaining at each router a table of the best known distance to each destination and the best link to reach it. In this context this article will focus on adaptive routing algorithms capable of simultaneously guaranteeing several QoS constraints such as end-to-end latency, jitter, cost, loss probability, bandwidth or any other measure of the quality of a particular route. The complexity associated with routing problems with multiple constraints is unaffordable to tackle with enumerative procedures even in its simplest version with two independent constraints [4]. In addition, an adaptive algorithm to efficiently handle multiple QoS constraints must be flexible enough to withstand fluctuations in any of the network parameters involved in such restrictions, as well as to accommodate traffic from new services that may compromise the performance figures of already established routes.

To alleviate the computational burden resulting from multiple simultaneous restrictions imposed over routing processes, a brief review of the related literature reveals that stochastic global search algorithms have been used extensively to optimize the discovery of feasible routes, with particular emphasis on algorithms inspired by ant colonies [5,6,7] and genetic algorithms [8,9,10] and references therein). This article joins this active line of research by analyzing the applicability and practical performance of other recently reported bio-inspired algorithms: Coral Reefs Optimization [13] and the Firefly algorithm [12], both featuring distinctive characteristics with respect to other evolutionary computation schemes. To the best of the authors, no previous reference of the application of these optimization meta-heuristics to network routing problems has been made public in the scientific literature. The manuscript will not only formulate the problem under consideration in a mathematically formal fashion, but will also describe thoroughly the solution encoding approach utilized to represent the successively refined routes and the operators driving the two solvers under comparison. Results from Monte Carlo simulations will be discussed to evince the performance merits of both schemes over networks of increasing complexity, specially in what regards to their speed of convergence.

2 Problem Formulation

The mathematical formulation of the aforementioned multiply constrained routing problem stems from the analogy of a communication network with a non-directed graph $G(\mathcal{V}, \mathcal{E})$, where $\mathcal{V}$ is the set of vertexes in the graph representing each of the network's constituent nodes; and $\mathcal{E}$ the group of edges in the graph that correspond to available links in the network. Each link $e \in \mathcal{E}$, defined between a source node u and a destination node v, is characterized by I coefficients or weights $\omega_i(e) \geq 0$ ($i \in \{1, \ldots, I\}$), serving as link's quality separate metric (e.g. cost, latency, available bandwidth, etc). Considering precedent definitions, multi constraint routing problem is expressed as finding a path $\mathbf{e} \triangleq \{e_1, \ldots, e_N\}$ (with $e_n \in \mathcal{E}$) from source node $s \in \mathcal{V}$ to destination node $t \in \mathcal{V}$ so that

$$W_i(\mathbf{e}) \triangleq \sum_{\forall e \in \mathbf{e}} \omega_i(e) \leq L_i, \quad i = 1, \ldots, I, \tag{1}$$

where $W_i(\mathbf{e})$ stands for i-th QoS measure's value for path $\mathbf{e}$, and $\{L_i\}_{i=1}^{I}$ are network operator enforced restrictions or QoS upper bounds. Notice that the problem hence formulated seeks disclosure of feasible, yet not necessarily optimal path in the sense of the QoS of the whole route (though its generalization in this sense is immediate). At the same time, it is important to stress that in its formulation (1) several routes may meet these restrictions simultaneously. If that were the case, and with no further criteria than the above stated, the interest lies in finding at least one of them.

In order to tackle the above problem by means of a fitness-driven solver, an objective function must be devised so as to guide the underlying optimization algorithm towards routes of increased optimality that eventually satisfy the set of imposed constraints. To this end the fitness of a certain route **e** is defined by

$$f(\mathbf{e}) = \max \left\{ \frac{W_1(\mathbf{e})}{L_1}, \frac{W_2(\mathbf{e})}{L_2}, \ldots, \frac{W_I(\mathbf{e})}{L_I} \right\}, \tag{2}$$

where following Expression (1) L_i represents i-th QoS parameter's maximum accepted value, and $W_i(p)$ is the aforestated parameter's total value along path **e**. Candidate paths will be ranked and arranged during the search procedure of the subsequently presented algorithms based on this fitness function: the lower the value of the fitness is, the closer path **e** to become a feasible route will be.

This formally defined objective function would be sufficient if all evaluated routes were valid. However, the route can be invalid if 1) a node along the route is connected to another node with which there is no real connection; 2) the path is short enough not to reach the destination; and/or 3) a node becomes part of the path repeatedly, forming a closed loop. In order to avoid unnecessary processing, in these cases the objective function (2) should be penalized so that the route is not eligible as optimal. This penalty will be set proportional to the number of times each of the aforementioned validity breaking circumstances holds for the route at hand, hence prioritizing those routes that are more likely to become valid if slight reparation procedures are included in the route optimization algorithm.

3   Proposed Route Optimization Algorithms

In general routing problems feature certain characteristics that hinder the adoption of population-based stochastic optimization techniques for their resolution:

1. In general, not all nodes are interconnected, so the initialization of the population of candidate solutions and its subsequent refinement require additional repairing methods, since not all randomly-generated solutions are feasible.
2. Due to this potentially sparse connectivity, operators associated with population-based optimization techniques must have been designed to operate with variable alphabets. In other words, procedures used for either diversification or intensification are less flexible and must be adapted in order to produce a feasible solution, as a necessary condition to reach an optimal solution.
3. In this family of problems, due to the repeated walk of the search process over one or more nodes, the probability of generating a loop on iteratively refined routes is very high. In order to prevent the existence of both finite and infinite loops, and taking into account that neither route where any node has been visited more than once should represent an optimum path, when it comes to initialize or improvising new routes the restriction that the route must not pass by any node more than once should be set.

It was not until the recent publication of the so-called MURUGA algorithm (*Multi-constraint QoS Unicast Routing Using Genetic Algorithm* [11]) that a routing algorithm able to efficiently withstand several quality of service constraints through a genetically inspired procedure was proposed. The algorithm uses genetic operators that operate on chromosomes that represent, by a binary encoding and a register of the directed neighborhood or "locus" of each one of the nodes, the candidate routes between considered source and destination. However, this proposal suffers from several shortcomings. To begin with, when ensuring a biunivocal identification of the route produced on the basis of its binary representation in the population, it is necessary to register the "locus" of all the nodes for the origin-destination pair at issue. In this context "locus" stands for the subset of neighboring nodes of a particular node and is calculated 1) sequentially node-by-node, starting at the origin node; and 2) recursively considering the "locus" calculated for previous nodes to avoid loops. Hence, the procedure for calculating locus is not naïve, with a complexity asymptotically comparable to the generation of all routes. Likewise, the simplicity of the coding in [11] has also an impact on the procedure for calculating the value of the metrics associated with generated routes. In the event that the metric takes the same value for all links, the calculation of the value of each of the metrics of the route will only depend on that value and the Hamming weight (i.e. number of ones) of its encoded representation, since the latter corresponds to the number of hops minus 2 of the route in question. Conversely, if the links are characterized by very divergent values of quality of service metrics, the algorithm must calculate them by walking through the structure of stored "locus", which implies a significant processing overhead.

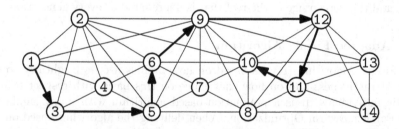

Figure 1. Schematic diagram of the linked route encoding scheme utilized in this work for a network with $N = 14$ nodes. The route $\{3, \Diamond, 5, \Diamond, 6, 9, \Diamond, \Diamond, 12, \Diamond, 10, 11, \Diamond, \Diamond\}$ from node 1 to node 10 is marked in bold line.

To address the above drawbacks and deficiencies of the MURUGA algorithm, this article resorts to a linked encoding approach for representing solutions in the multi-constrained routing problem at hand. A route from source node $s \in \mathcal{V}$ to destination node $t \in \mathcal{V}$ through a network with N nodes will be represented by a list of integers $\{x_1, x_2, \ldots, x_N\}$ such that x_n will denote the index of the node following node n in the route from s to t. For instance, if $N = 10$ the route $2 \rightarrow 5 \rightarrow 10 \rightarrow 9 \rightarrow 7$ from $s = 2$ to $t = 7$ will be represented by

$\{\Diamond, 5, \Diamond, \Diamond, 10, \Diamond, \Diamond, \Diamond, 7, 9\}$, where $\Diamond$ denotes "any value" as the values in these positions do not alter the value of the fitness.

Two algorithms will be used to solve the problem: Coral Reefs Algorithm and Firefly Algorithm. In this manner it will be possible to assess whether both algorithms are suitable for overcoming this issue, and whether any of them outperforms the other.

3.1 Coral Reefs Optimization

The Coral Reefs Optimization (CRO) algorithm is the first meta-heuristic solver that will be utilized so as to lead a population of network routes towards regions of progressively enhanced optimality. First proposed in [13], this solver emulates the observed biological behavior of coral reefs in different stages: formation, reproduction, competition for space in the reef by newly produced larvae and depredation by other animals. The algorithm constructs and refines a reef of solutions to a given optimization problem (corals) by imitating the way corals reproduce in Nature (i.e. by implementing broadcast spawning, brooding and budding operators). CRO has indeed conceptual similarities to Evolutionary Algorithms and Simulated Annealing, and has been shown to outperform other meta-heuristic approaches in different scenarios.

One of the most acknowledged drawbacks of this algorithm is the relatively high number of parameters that drive its search process: the percentage of already allocated positions on the reef (r_0); the percentage of larvae subject to crossover (F_b), i.e. $1 - Fb$ be the percentage of larvae over which the mutation will be applied; the number of attempts allowed for larvae to deploy on the reef (κ); the budding percentage, i.e. the percentage of best solutions to be replicated (F_a); and the percentage of the reef that is depredated at every generation (F_d).

3.2 Adapted Firefly Algorithm

The Firefly Algorithm (FA) is based on how fireflies use its brightness to gain more attractiveness and on how they move on this basis with respect to other nearby individuals. It is a population-based algorithm with many similarities with Particle Swarm Optimization. When defining the algorithm based on natural firefly behavior three assumptions are set [12]:

1. There is no sex distinction for fireflies.
2. The attractiveness is proportional to the brightness, and both decrease as distance increases. If there are two fireflies, the less brighter one will move towards the brightest one. If there is no other firefly around, the firefly will move randomly. To simulate random movement each of the dimensions (positions) of the firefly will be changed under a probability p_r.
3. A firefly's brightness is given by objective function's value, i.e. the better the value of the objective function is, the higher its brightness will be.

Attraction among fireflies is set based on the perceived light intensity at a distance r, which is driven by a light absorption rate γ as $I(r) = I_0 e^{-\gamma r^2}$, with I_0

denoting the light intensity of the emitting firefly. Based on this distance-based degradation, the update for any pair of fireflies $\mathbf{x}_i$ and $\mathbf{x}_j$ is given by

$$\mathbf{x}_i^{t+1} = \mathbf{x}_i^t + \beta \exp[-\gamma r_{ij}^2](\mathbf{x}_j^t - \mathbf{x}_i^t) + \alpha\epsilon, \tag{3}$$

where α is a step size and ϵ denotes the realization of a multi-dimensional random variable drawn from a Gaussian or any other distribution. It should be noted that the above formulae assumes real-valued solutions. In order to adapt the algorithm to discrete optimization problems the inner concepts of *movement*, *distance* and *absorption* must be redefined:

– The movement from one firefly (solution) to another will be measured in terms of the number of nodes in the attracting firefly are replicated in the attracted firefly.
– Likewise, distance $d_{i,j}$ between fireflies i and j will be expressed as the number of nodes in which they differ.
– The effective brightness perceived by the attracted firefly will still be driven by parameter γ, and will involve replicating a fraction θ of the compounding entries of the attracting firefly. Since the algorithm is expected to be more explorative the lower this parameter is, the fraction of replicated positions from firefly i to j will be given by $\theta = d_{i,j}(1 - \gamma)/(N - 1)$. Only those positions representing valid nodes (i.e. not equal to $\Diamond$) and located between source and destination nodes will be replicated.

3.3 Route Generation: Repair Procedure

As has been previously mentioned the criterion that routes that pass through any node more than once can not led to optimum will apply to route generation. Therefore, if a n node is connected to a single node n', then node n' will never be used as intermediate node of the path and therefore will not be part of the route, except if and only if it is also the destination node. To avoid this type of situation, i.e., the existence of nodes that do not pay service (or do so only rarely), an additional restriction is established for network generation: all nodes must be connected at least with other two nodes.

There are several ways to initialize candidate solutions: in this study the considered optimization algorithms operate a population of valid[3] solutions uniformly selected over the complete set of existent valid solutions. This approach asks for prior generation of all routes between source and destination nodes. Although it improves algorithm's convergence speed to an optimal solution (since all produced solutions are valid), this option suffers from a great disadvantage: it is necessary to generate – not evaluate – all routes, which in highly complex networks may yield a complexity at the same level as that of an exhaustive search technique. Considering that a fully interconnected network of N nodes, the number of routes between whatever 2 nodes is given by $\sum_{k=2}^{N}(N-2)!/(N-k)!$, which for e.g. $N = 13$ amounts up to approximately 108 million routes.

---

[3] Valid routes are those that do no violate the connectivity of the network, whereas feasible routes are those valid solutions that also comply with the constraints.

Since the algorithmic scope will be placed in population-based heuristics, M candidate solutions are generated uniformly at random so as to produce the initial population. This random generation procedure comes at the risk of including invalid entries in the population. To minimize (yet not necessarily overcome) this risk, solutions can be repaired right after they have been produced either during initialization or as a result of the evolutionary operators that characterize the meta-heuristic algorithms explained in the next subsections. When a given generated route is not valid at a certain node, the algorithm will attempt at reconnecting its preceding node to any other in range. In case there is no other node within coverage the last node of the route is marked with a special flag that denotes its ineligibility as an intermediate node of a route. The experimental benchmark later discussed in the article will include such algorithms *with* and *without* repairing stage so as to quantify its impact on the search convergence.

4 Simulation Results

Computer simulations have been run so as to shed light on the comparative performance of the two proposed heuristics and to assess the impact of the repair procedure on their search efficiency. To this end 10 different networks have been created by deploying $N \in \{50, 100, 150\}$ nodes uniformly at random over a square grid of size 100×100. A binary coverage model is imposed for establishing the connectivity among nodes, with adjusted radii so as to ensure that in all networks nodes are connected to at least 10 neighboring nodes. The source node is always the one labeled as 1, whereas the destination node is forced to be the furthest to 1. Performance scores will be averaged over 10 Monte Carlo realizations of each of 4 different routing schemes over every single simulated network: CRO and FA without (CRO, FA) and with (CRO-R, FA-R) repair procedure. For the sake of fairness in the comparison the population size (i.e. number of corals in CRO and number of fireflies in FA) and the maximum number of iterations is set to $M = 80$ and $\mathcal{I} = 300$. Parameters of all techniques in the benchmark have been optimized in off-line simulations (not shown for brevity).

Two different QoS measures will be taken into account: end-to-end delay $(W_1(\mathbf{e}))$ and the number of hops $(W_2(\mathbf{e}))$. On one hand, under the ITU-T G.1010 recommendation [14] the maximum admissible delay for a route depends roughly on the type of network traffic it conveys. The upper bound $(L_1 = 0.15$ seconds) selected for the simulation corresponds to video streaming. Without loss of generality the delay of link $e_{i,j}$ will depend proportionally on the distance $d_{i,j}$ between its connected nodes i and j (hence modeling the propagation delay) and the overall number of nodes N as $\omega_2(e_{i,j}) = 0.2L_1 d_{i,j}/N$. On the other hand, the maximum number of hops for the route will be bounded by $L_2 \in \{3, 5\}$ hops.

The results obtained for the benchmark are collected in Table 1 in the form of average number of iterations until a feasible/valid solution is met $(\mathcal{I}_{avg})$, fraction of Monte Carlo experiments where the algorithm at hand has met a feasible solution $(|\mathbf{x}|_{\checkmark})$, and in-feasibility share of the unfeasible solutions $(|\mathbf{x}|_{\times} = A/B$, where A and B stand for the percentage of Monte Carlo experiments not having met the delay and maximum hops targets, respectively). It should be clear that $100 - (|\mathbf{x}|_\checkmark + A + B - A \cap B)$ quantifies the number of Monte Carlo realizations that have not produced any route – neither valid nor feasible – from source to destination, with $A \cap B$ denoting the percentage of experiments having failed to meet both targets. A first glance at the results reveals that the worst performing approach is CRO, which is outperformed by FA for all network sizes due to its more suited design to the problem at hand. Remarkably it is when the repair process is incorporated to the CRO search thread when the resulting CRO-R solver becomes competitive in the benchmark in terms of valid routes. On the other hand, when $N = 50$ it can be seen that the delay constraint impacts more noticeably in the share of unfeasible routes due to the fact that since nodes are forced to be connected to 10 neighboring counterparts for all networks (which makes it more likely to get loops within the route) and the increased average distance between nodes. Furthermore, in this same case the speed of convergence for FA, FA-R and CRO-R (given by $\mathcal{I}_{avg}$) is higher due to the larger density of possible routes.

Table 1. Performance statistics of the compared algorithms.

		$N = 50$		$N = 100$		$N = 150$	
		$L_2 = 3$	$L_2 = 5$	$L_2 = 3$	$L_2 = 5$	$L_2 = 3$	$L_2 = 5$
CRO	$\mathcal{I}_{avg}$	35.79	50.26	28.07	47.00	83.55	110.85
	$|\mathbf{x}|_\checkmark$	48%	58%	24%	47%	7%	11%
	$|\mathbf{x}|_\times$	39%/9%	33%/0%	3%/20%	4%/7%	0%/1%	1%/2%
FA	$\mathcal{I}_{avg}$	1.28	3.71	4.47	4.36	27.00	14.10
	$|\mathbf{x}|_\checkmark$	90%	90%	100%	100%	98%	100%
	$|\mathbf{x}|_\times$	10%/0%	10%/0%	0%/0%	0%/0%	0%/2%	0%/0%
CRO-R	$\mathcal{I}_{avg}$	2.55	14.64	5.06	1.49	29.01	2.76
	$|\mathbf{x}|_\checkmark$	90%	85%	99%	100%	96%	100%
	$|\mathbf{x}|_\times$	10%/0%	15%/0%	0%/1%	0%/0%	0%/4%	0%/0%
FA-R	$\mathcal{I}_{avg}$	1.10	2.83	2.68	1.11	13.67	1.42
	$|\mathbf{x}|_\checkmark$	90%	90%	100%	100%	100%	100%
	$|\mathbf{x}|_\times$	10%/0%	10%/0%	0%/0%	0%/0%	0%/0%	0%/0%

5   Conclusions and Future Work

This manuscript has presented a general bio-inspired heuristic framework for multiply-constrained network routing problems. In this general class of optimization paradigms the design goal is not strictly focused on optimizing a given fitness function, but rather to find at least one route between any given source-destination pair that fulfills manifold QoS constraints. This work has elaborated on the formal definition of the problem and the need for including repair methods within the search procedure so as to ensure the validity of successively evolved solutions to the problem. The framework has been put to practice by using two recently proposed bio-inspired solvers (CRO and a modified FA for discrete optimization), whose comparative performance has been analyzed in several synthetic networks of diverse size. The obtained results evince that the utility of the proposed framework is promising in the context of QoS-based network routing, and paves the way for future research aimed at verifying whether this potentiality holds in large-scale networks, possibly by incorporating distributed versions of its compounding heuristics.

6 Acknowledgements

This work was supported by the Basque Government under its ELKARTEK research program (ref: KK-2015/0000080, BID3A project).

References

1. Baransel, C., Dobosiewicz, W., Gburzynski, P.: Routing in Multihop Packet Switching Networks: Gb/s Challenge. IEEE Network, 38–60 (1995)
2. ITU-T Recommendation: E.800: Terms and Definitions Related to Quality of Service and Network Performance including Dependability (1994)
3. Vetriselvan, V., Patil, P. R., Mahendran, M.: Survey on the RIP, OSPF, EIGRP Routing Protocols. International Journal of Computer Science and Information Technologies, 5(2), 1058-1065 (2014)
4. Wang, Z., Crowcroft, J.: Quality of Service Routing for Supporting Multimedia Applications. IEEE Journal on Selected Areas in Comm., 14, 7, 1228–1234 (1996)
5. Di Caro, G.: Ant Colony Optimization and its Application to Adaptive Routing in Telecommunication Networks. PhD Thesis, Universite Libre de Bruxelles (2004)
6. Sandalidis, H. G., Mavromoustakis, C. X., Stavroulakis, P. P.: Performance Measures of an Ant based Decentralised Routing Scheme for Circuit Switching Communication Networks. Soft Computing, 5, 4, 313–317 (2001)
7. Wedde, H. F., Farooq, M.: Beehive: New Ideas for Developing Routing Algorithms inspired by Honey Bee Behavior. Handbook of Bioinspired Algorithms and Applications, 321–339 (2005)
8. Nagib, G., Ali, W. G.: Network Routing Protocol using Genetic Algorithms. International Journal of Electrical & Computer Sciences, 10, 2, 40–44 (2010)
9. Ahn, C. W., Ramakrishna, R. S.: A Genetic Algorithm for Shortest Path Routing Problem and the Sizing of Populations. IEEE Transactions on Evolutionary Computation, 6, 6, 566–579 (2002)
10. Wedde, H. F., Farooq, M.: A Comprehensive Review of Nature Inspired Routing Algorithms for Fixed Telecommunication Networks. Journal of Systems Architecture, 52, 461–484, (2006)
11. Leela, R., Thanulekshmi, N., Selvakumar, S.: Multi-Constraint Qos Unicast Routing Using Genetic Algorithm (MURUGA). Applied Soft Computing, 11, 2, 1753–1761 (2011)
12. Yang, X.-S., He, X.: Firefly Algorithm: Recent Advances and Applications. International Journal of Swarm Intelligence, 1, 1, 36-50, 2013.
13. Salcedo Sanz, S., Del Ser, J., Landa-Torres, I., Gil-Lopez, S., Portilla-Figueras, J. A.: The Coral Reefs Optimization Algorithm: A Novel Metaheuristic for Efficiently Solving Optimization Problems. The Scientific World Journal, 739768 (2014)
14. ITU-T G.1010 Recomendation: End-user Multimedia QoS Categories (2002)

Dealing with the Best Attachment Problem via Heuristics

M. Buzzanca, V. Carchiolo, A. Longheu, M. Malgeri, and G. Mangioni

Abstract Ordering nodes by rank is a benchmark used in several contexts, from recommendation-based trust networks to e-commerce, search engines and websites ranking. In these scenarios, the node rank depends on the set of links the node establishes, hence it becomes important to choose appropriately the nodes to connect to. The problem of finding which nodes to connect to in order to achieve the best possible rank is known as the *best attachment problem*. Since in the general case the best attachment problem is NP-hard, in this work we propose heuristics that produce near-optimal results while being computable in polynomial time; simulations on different networks show that our proposals preserve both effectiveness and feasibility in obtaining the best rank.

1 Introduction

The ranking of nodes within a network emerged during last years in several scenarios, from trust-based recommendation networks [1, 2] to website relevance score in search engines [3, 4], e-commerce B2C and C2C transactions [5, 6].

Within each specific framework, different proposals exist about the meaning of nodes (agents, peer, users) and about the rank assessment algorithm; in all of them, the higher is the rank of a node, the higher is its legitimacy. The rank depends on the set of out-links each node establishes with others [7], and on the set of in-links it receives.

Links model trust in trust networks, hyperlinks in website ranking, or buyer-seller relationships in the e-commerce context. The consequence is that whenever a new node joins the network, it attempts to increase its rank by selecting a set of existing nodes to be pointed by a couple of question arise: the problem of finding the nodes to attach to in order to achieve the best possible rank is known as the *best attachment* problem. Given that the network is usually modeled as a directed graph

---

Dip. Ingegneria Elettrica, Elettronica e Informatica - Università degli Studi di Catania - Italy

© Springer International Publishing AG 2017
C. Badica et al. (eds.), *Intelligent Distributed Computing X*,
Studies in Computational Intelligence 678, DOI 10.1007/978-3-319-48829-5_20

$G(V,E)$, finding the k best attachments for a given node $i \in V$ consists in finding a set $S \subseteq S_{i,k}$ that maximizes the rank P_i by changing its in_i to $in_i \cup S$, where $in_i = \{v \in V : \exists e(v,i) \in E\}$ and $S_{i,k} = \{S \in V : |S| = k, S \cap in_i = \emptyset\}$. Essentially, we need to establish k links from the k nodes that will improve i's rank value up to the highest [8]

While the best attachment problem definition we provided is general, our analysis is focused on the well-known PageRank [9] algorithm and on its properties. Intuitively, we may think that if we select the first k nodes in the ordered Pagerank vector we would reach the optimal solution to the problem, but this is not usually the case. The PageRank algorithm is driven by the node backlinks, not forward links. This means that even if i connects to an highly ranked node, unless that node points towards i as well, it is not guaranteed that its Pagerank is positively affected: depending on the network topology, it may even be possible that its Pagerank can decrease. In general, the best attachment problem is NP-hard, as it would require to compute $\binom{k}{n}$ Pageranks; as analytically demonstrated in [10], the problem is actually W[1]-hard, making it unfeasible to compute the optimal solution in real-life scenarios, as even with small networks we would need to compute the Pagerank millions of times. Because of these computational issues, it is necessary to find an approximation algorithm to choose a solution acceptably close to the optimum in a polynomial time.

In [10] however authors also show that there exist both upper and lower bounds for certain classes of heuristics. It is not always possible to calculate these bounds as sometimes the computational complexity of these calculations is NP-hard as well, nevertheless finding bounds of heuristics lets us rank their theoretical accuracy.

In this paper we propose several heuristics that aim at providing a near-optimal solution in reasonable time, preserving effectiveness while achieving practical feasibility. We applied the proposed heuristics to different syntetic networks by simulation, and we compared the results.

In our previous studies [11, 12, 13], we experimented with a brute-force solution to the best attachment problem, and we analyzed the cost of link building and its dynamic. There is however a number of works in literature concerning the best attachment problem. In [14], authors use asymptotic analysis to see how a page can control its pagerank by creating new links. A generalization of this strategy to websites with multiple pages is described in [15]. In [16] authors model the link building problem by using constrained Markov decision processes. In [17] author demonstrates that by appropriately changing node outlinks the resulting PageRank can be dramatically changed. A related problem of the best attachment is the pagerank updating computation issue, described in [18] and expanded upon in [19] among a comperhensive list of pagerank related challenges. An heuristic to find the first eigenvalue of the transition matrix is introduced in [20].

The paper is organized as follows. In section 2 we illustrate the set of heuristics we considered, whereas in section 3 we show simulations for each approach, together with heuristics comparisons. concluding remarks and future works are briefly discussed in section 4.

2 Heuristics

As discussed, the problem to find the best set of nodes allowing us to gain the best reputation is not feasible due to computability complexity, therefore we propose some heuristics to approximate the solution. In this section, we detail the algorithms and discuss both pros and cons while in the next section we show some experiments that highlight the behavior of the proposed heuristics and their dependence on network topology.

The main goal of the heuristics is to allow a new node (called *me*), to find a trade-off between the minimization of steps, the cost of new links creration, and the rank position.

Note that the cost of creating a link is both the computational effort needed to evaluate the increasing of rank (if any) and the cost needed to prevail on a node $(x \in V)$ to create a link with *me*.

The computational complexity of all stategies depends on the computational complexity of pagerank evaluation, in the following called $\mathscr{O}(PR)$. Using the Gauss method it would require $\mathscr{O}(|V|^3)$, however using iterative approximation it would require $\mathscr{O}(PR) = m * |E|$, where m is the number of iterations needed to get a good approximation [18].

We discuss in the following subsections some heuristiscs based both on naive approach and on PageRank evaluation aiming at reducing the complexity in solving the best attachment problem.

2.1 Naive Algorithms

The simplest approach to get best attachment problem consists in selecting the nodes k to populate S randomly until we get our goal. Of course, this strategy is quite naive but it is simple to implement and the cost to create link is proportianal to number of links only. The computational complexity depends only on the number of steps *me* needs to get the best position, i.e. $\mathscr{O}(random) = \mathscr{O}(m)$ where m is the number of steps to converge. The algorithm is depicted in Algorithm 1. Of course *Random*

Algorithm 1 Random Choice Algorithm

1: **procedure** RANDOM(V, me)   ▷ V is the set of vertices, *me* is the target node
2:     $T = V$
3:     $S = \emptyset$
4:     **while** $rank_{me} > 1$ **do**   ▷ Iterate until rank is the best
5:         random_select $x \in T$
6:         $T = T - \{x\}$
7:         $S = S \cup \{x\}$
8:     **end while**
9: **return** S
10: **end procedure**

---

strategy is quite trivial and does not rely on any of the network's property. However when the topology is almost regular and the distribution of both in- and out-degree- is also regular, this algorithm should perform as well as others more complex.

Another simple approach to find a node to be pointed by comes from the degree of target node *me*, so we can use in-, out- or full-degree of the target node as a selection criteria. The algorithm proposed is reported in Algorithm 2.

Algorithm 2 Degree Algorithm

1: **procedure** DEGREE(V, me)   ▷ V is the set of vertices, *me* is the target node
2:    $T = V$
3:    $S = \emptyset$
4:    **while** $rank_{me} > 1$ **do**   ▷ Iterate until rank is the best
5:       select $x \in T$, where degree(x) is max   ▷ x node with highest degree in T
6:       $T = T - \{x\}$
7:       $S = S \cup \{x\}$
8:    **end while**
9: **return** S
10: **end procedure**

Let us note that the complexity of algorithm is $\mathcal{O}(Degree) = \mathcal{O}(m) * \mathcal{O}(select)$, where m is the number of steps and $\mathcal{O}(select)$ is the complexity to find the node having maximum in-, out- or full-degree. This last term depends on the type of data structure used to store T.

2.2 Pagerank Based Algorithms

While *Random* approach - being a trivial one - uses no information about network topology and nodes characteristics, the next strategies aim at overcoming this limit using information about the ranking of nodes increasing as little as possible the computability complexity. Since higher rank nodes are the most popular inside the networks we select the in-link node according to its reputation. However, change of topology due to the new connection could affect the ranking of the nodes, therefore we can outline different strategies, depending on the how frequently ranking is evaluated, as reported below.

1. *Anticipated Rank* strategy: the rank is calculated just before starting the search and used throughout the whole algorithm to select the (not used) in-link node. Based on the way the ranking is computed we can distinguish two algorithms:

   - *Anticipated_value* (Algorithm 3): the ranking is calculated by ordering nodes according to the node's pagerank value; the idea here is to select first nodes with higher pagerank value.
   - *Anticipated_outdeg* (Algorithm 4): the ranking is computed by ordering nodes according the ratio between node's pagerank value and node's out degree; in

this approach, we first select the node that *"transfers"* the highest pagerank value to its out-neighbourhood.

2. *Current Rank* strategy (Algorithm 5): the rank is recalculated each iteration and the best not used node is selected. We always use current rank.
3. *Future Rank* strategy (Algorithm 6): algorithm evaluates all possible connection - one step behind - and select the connection giving *me* the best rank. This strategy should be the more effective but it is the more expensive.

Algorithm 3 Anticipated_value Algorithm

1: **procedure** ANTICIPATED_VALUE(V, me)   ▷ V is the set of vertices, *me* is the target node
2:   $S = \emptyset$
3:   $R = pagerank$ ▷ R is the set of $n|n \in V$ ordered according to their pagerank value
4:   **repeat**
5:     $x = first(R)|x \notin S$
6:     $R = R - \{x\}$
7:     $S = S \cup \{x\}$
8:   **until** $(rank_{me} = 1)$ ▷ Node will always get best position before R became empty
9: **return** S
10: **end procedure**

Algorithm 4 Anticipated_outdeg Algorithm

1: **procedure** ANTICIPATED_OUTDEG(V, me)   ▷ V is the set of vertices, *me* is the target node
2:   $S = \emptyset$
3:   $R = pagerank/out\_degree$ ▷ R is the set of $n|n \in V$ ordered according to the ratio pagerank / degree
4:   **repeat**
5:     $x = first(R)|x \notin S$
6:     $R = R - \{x\}$
7:     $S = S \cup \{x\}$
8:   **until** $(rank_{me} = 1)$ ▷ Node will always get best position before R became empty
9: **return** S
10: **end procedure**

The complexity of the strategy *Anticipated_value* and *Anticipated_outdeg* is almost the same of *Degree* strategy, in fact PageRank is evaluated only once before starting the iteration. However the more nodes are selected, the more often network topology changes therefore the results of initial PageRank evaluation become less accurate.

Current is the second strategy based on pagerank we propose in this work; it selects the node according to its rank, but re-evaluate pagerank at each iteration. The complexity of this algorithm is $\mathcal{O}(current) = \mathcal{O}(m) * max[\mathcal{O}(select), \mathcal{O}(PR)]$ and it is quite higher than all previous algorithms, since generally $\mathcal{O}(PR)$ is higher than $\mathcal{O}(select)$. However, this algorithm seems to capture the network dynamics due to creation of new links better than previous ones.

Algorithm 5 Current Algorithm

1: **procedure** CURRENT(V, me)   ▷ V is the set of vertices, me is the target node
2: $\quad S = \emptyset$
3: $\quad$ **repeat**
4: $\quad\quad$ compute pagerank
5: $\quad\quad$ select $x \in V | x \notin S$ where pagerank is max
6: $\quad\quad S = S \cup \{x\}$
7: $\quad\quad E = E \cup (x, me)$ ▷ Connect x to me
8: $\quad$ **until** ($rank_{me} = 1$)
9: **return** S
10: **end procedure**

Future is the last algorithm proposed; it tries to evaluate the best node to be pointed by via evaluating the PageRank that will be obtained by *me* after the arc creation. This algorithm needs a continue re-evaluation of pagerank and its complexity is $\mathcal{O}(future) = \mathcal{O}(m) * max[\mathcal{O}(select), \mathcal{O}(PR) * \mathcal{O}(|V|)]$, that can be rewritten as: $\mathcal{O}(future) = \mathcal{O}(m) * \mathcal{O}(PR) * \mathcal{O}(|V|)$ At first glance, this algorithm seems to

---

Algorithm 6 Future Algorithm

1: **procedure** FUTURE(V, me)   ▷ V is the set of vertices, me is the target node
2: $\quad S = \emptyset$
3: $\quad$ **repeat**
4: $\quad\quad R = \emptyset$
5: $\quad\quad$ **repeat**
6: $\quad\quad\quad$ connect node $x \in V | (x \notin R) \, and \, (x \notin S)$ to node me
7: $\quad\quad\quad$ calculate pagerank
8: $\quad\quad\quad R = R \cup \{(x, rank_{me}^{x})\}$
9: $\quad\quad\quad$ disconnect x from me
10: $\quad\quad$ **until** $|R| = |V|$ ▷ Iterate over all nodes belonging to V
11: $\quad\quad$ select $(x, rank_{me}^{x}) \in R$ where $rank_{me}^{x}$ is max
12: $\quad\quad S = S \cup \{x\}$
13: $\quad\quad E = E \cup (x, me)$ ▷ Connect x to me
14: $\quad$ **until** ($rank_{me} = 1$)
15: **return** S
16: **end procedure**

---

be optimal since it selects the node which gives the best rank, but its complexity is higher than all previous algorithms. To partially overcome this problem we propose another heuristic that does not iterate over all nodes but selects randomly N nodes to which apply the *Future Algorithm* approach.

3 Results

To study the performance of the heuristics proposed in the previous section, we conduct a set of experiments on two well-known family of networks. In particular, we consider Erdos–Renyi random networks (ER) and scale–free (SF) networks.

A random ER network is generated by connecting nodes with a given probability p. The obtained network exhibit a normal degree distribution [21]. A scale–free network (SF) [22] is a network whose degree distribution follows a power law, i.e. the fraction $P(k)$ of nodes having degree k goes as $P(k) \sim k^{-\gamma}$, where γ is typically in the range $2 < \gamma < 3$. A scale–free network is characterized by the presence of *hub* nodes, i.e. with a degree that is much higher than the average. The scale-free network employed in this work is generated by using the algorithm proposed in [23] as implemented in the Pajek[24] tool.

Simulations have been performed by using 100k nodes networks of both topologies. Table 1 reports the main topological properties of such networks.

Table 1 Networks topological parameters

name	#nodes	#links	average degree
ER	100000	1401447	28.028
SF	100000	1394248	27.884

In figure 1 the degree distributions of 100K nodes networks is shown. As expected, the ER network (figure 1(a)) exhibits a normal degree distribution, while the SF degree distribution (figure 1(b)) follows a power-law.

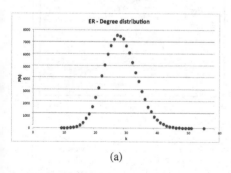

(a)

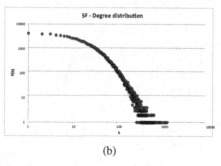

(b)

Fig. 1 Degree distribution for ER and SF networks

Figure 2 reports the rank of *me* with respect to the number of in-links and steps for all the algorithms presented in the previous section. The best rank is represented by the position 1, so at the beginning *me* has the worst rank, i.e. 100001. The figure shows that *Degree_out* and *Degree_full* are the worst algorithms in terms of performance. *Random, Degree_in, Anticipated_value* and *Current* exhibits comparable performance. Despite the fact *Random* has the lower computational complexity among the proposed algorithms, it performs as well as more complex algorithms. *Future* and *Anticipated_outdeg* are the best algorithms when applied to ER networks. Surprisingly *Anticipated_outdeg* performs even better than *Future*, mainly during the initial part of the attachment process. In fact, as detailed in the figure inset, *Anticipated_outdeg* permits *me* to rapidly achieve a good rank, even if Future outperforms it after the step number 10. Let's note, however, that the computational complexity of *Anticipated_outdeg* is far less than *Future*, making the application to very large networks feasible.

In the figure 3 the simulation results for SF network is shown. It's clear that *Random, Anticipated_outdeg* and *Future* outperform the other algorithms proposed in the previous section. As in the case of ER networks, the performance of the *Anticipated_outdeg* algorithm is surprising since its trend is comparable to the more complex algorithm *Future*. In addition, *Anticipated_outdeg* allows the node *me* to reach the best rank in only 18 steps against the 36 required by *Future*. On the other hand, *Random* algorithm gets the best rank in 69 steps, that is a very good figure considering the size of the network (100K nodes) and the random selection strategy.

An additional note concerns the comparison of the algorithms in the two different network topologies. In general, in SF network algorithms tend to gain rapidly a good positioning than in ER network. As an example, *Future* permits the node *me* to gain the position number 29 in just 10 steps, whilst the same algorithm applied to ER reaches position 5966. On the other hand, in SF it is more difficult to obtain the rank 1, in fact in *Future* it happens at the step 36, while in ER the same algorithm

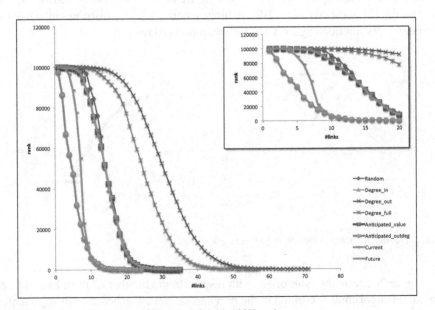

Fig. 2 Heuristics performance on ER network with 100K nodes

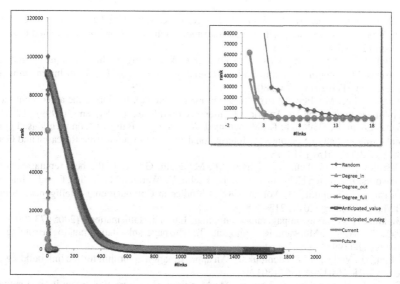

Fig. 3 Heuristics performance on SF network with 100K nodes

gains the rank 1 in 17 steps. This behaviour is probably due to the presence in SF networks of hubs and authority, that play a central role on the dynamic underlying the PageRank evaluation.

4 Conclusion

The problem of finding the in-neighbourhood of a target node to get its best pagerank is known as the best attachment or linking building problem. It is also known that it is very hard to solve this problem in a exact way. For this reason in this paper we propose a set of algorithms aiming at finding an approximated solution. They are mostly based on intuitive reasoning about the way pagerank works, ranging from a very simple random choice to a more sophisticated approach. All the proposed algorithms have been evaluated on Erdos-Renyi and scale–free networks of 100K nodes. Results confirm that simple algorithms, such as *Random* and *Anticipated_outdeg*, can exhibits performance comparable and, sometimes, better than more complex algorithms, thus permitting their application to analyze very huge networks in a reasonable time.

References

1. Weng, J., Miao, C., Goh, A., Shen, Z., Gay, R.: Trust-based agent community for collaborative recommendation. In: AAMAS '06: Proceedings of the fifth international joint conference on Autonomous agents and multiagent systems, New York, NY, USA, ACM (2006) 1260–1262
2. Liu, X.: Towards context-aware social recommendation via trust networks. In Lin, X., Manolopoulos, Y., Srivastava, D., Huang, G., eds.: Web Information Systems Engineering - WISE 2013. Volume 8180 of Lecture Notes in Computer Science. Springer Berlin Heidelberg (2013) 121–134

3. Pan, B., Hembrooke, H., Joachims, T., Lorigo, L., Gay, G., Granka, L.: In google we trust: Users decisions on rank, position, and relevance. Journal of Computer-Mediated Communication **12**(3) (2007) 801–823
4. Chauhan, V., Jaiswal, A., Khan, J.: Web page ranking using machine learning approach. In: Advanced Computing Communication Technologies (ACCT), 2015 Fifth International Conference on. (Feb 2015) 575–580
5. Fung, R., Lee, M.: Ec-trust (trust in electronic commerce): Exploring the antecedent factors. In: Proceedings of the 5th Americas Conference on Information Systems. (1999) 517–519
6. Sameerkhan, P., Shahrukh, K., Mohammad, A., Amir, A., Bali, A.: Comment based grading and rating system in e-commerce. International Journal of Engineering Research and General Science **3**(1) (2015) 1319–1322
7. Carchiolo, V., Longheu, A., Malgeri, M., Mangioni, G.: Gain the best reputation in trust networks. In Brazier, F., Nieuwenhuis, K., Pavlin, G., Warnier, M., Badica, C., eds.: Intelligent Distributed Computing V. Volume 382 of Studies in Computational Intelligence. Springer Berlin Heidelberg (2012) 213–218
8. Berkhin, P.: A survey on pagerank computing. Internet Mathematics **2** (2005) 73–120
9. Page, L., Brin, S., Motwani, R., Winograd, T.: The pagerank citation ranking: Bringing order to the web (1998)
10. Olsen, M., Viglas, A., Zvedeniouk, I.: An approximation algorithm for the link building problem. CoRR **abs/1204.1369** (2012)
11. Carchiolo, V., Longheu, A., Malgeri, M., Mangioni, G.: Users attachment in trust networks: Reputation vs. effort. Int. J. Bio-Inspired Comput. **5**(4) (July 2013) 199–209
12. Carchiolo, V., Longheu, A., Malgeri, M., Mangioni, G.: A heuristic to explore trust networks dynamics. In Zavoral, F., Jung, J.J., Badica, C., eds.: Intelligent Distributed Computing VII. Volume 511 of Studies in Computational Intelligence. Springer International Publishing (2014) 67–76
13. Carchiolo, V., Longheu, A., Malgeri, M., Mangioni, G.: The cost of trust in the dynamics of best attachment. Computing & Informatics **34**(1) (2015)
14. Avrachenkov, K., Litvak, N.: The effect of new links on google pagerank. Stochastic Models **22**(2) (2006) 319–331
15. de Kerchove, C., Ninove, L., Dooren, P.V.: Maximizing pagerank via outlinks. CoRR **abs/0711.2867** (2007)
16. Fercoq, O., Akian, M., Bouhtou, M., Gaubert, S.: Ergodic control and polyhedral approaches to pagerank optimization. IEEE Trans. Automat. Contr. **58**(1) (2013) 134–148
17. Sydow, M.: Can one out-link change your pagerank? In Szczepaniak, P.S., Kacprzyk, J., Niewiadomski, A., eds.: AWIC. Volume 3528 of Lecture Notes in Computer Science., Springer (2005) 408–414
18. Bianchini, M., Gori, M., Scarselli, F.: Inside pagerank. ACM Trans. Internet Technol. **5**(1) (Febraury 2005) 92–128
19. Langville, A., Meyer, C.: Deeper inside pagerank. Internet Mathematics **1**(3) (2004) 335–380
20. Nazin, A., Polyak, B.: Adaptive randomized algorithm for finding eigenvector of stochastic matrix with application to pagerank. In: Proceedings of the 48th IEEE Conference on Decision and Control, CDC/CCC 2009. (Dec 2009) 127–132
21. Albert, R., Barabasi, A.L.: Statistical mechanics of complex networks. Reviews of Modern Physics **74** (2002) 47
22. Barabasi, A.L., Albert, R.: Emergence of scaling in random networks. Science **286** (1999) 509
23. Pennock, D.M., Flake, G.W., Lawrence, S., Glover, E.J., Giles, C.L.: Winners don't take all: Characterizing the competition for links on the web. Proceedings of the National Academy of Sciences **99**(8) (2002) 5207–5211
24. Batagelj, V., Mrvar, A.: Pajek - program for large network analysis. (1999)

Part VII

Data Management

Towards Collaborative Sensing using Dynamic Intelligent Virtual Sensors *

Radu-Casian Mihailescu, Jan Persson, Paul Davidsson, Ulrik Eklund

Abstract The recent advent of 'Internet of Things' technologies is set to bring about a plethora of heterogeneous data sources to our immediate environment. In this work, we put forward a novel concept of dynamic intelligent virtual sensors (DIVS) in order to support the creation of services designed to tackle complex problems based on reasoning about various types of data. While in most of works presented in the literature virtual sensors are concerned with homogeneous data and/or static aggregation of data sources, we define DIVS to integrate heterogeneous and distributed sensors in a dynamic manner. This paper illustrates how to design and build such systems based on a smart building case study. Moreover, we propose a versatile framework that supports collaboration between DIVS, via a semantics-empowered search heuristic, aimed towards improving their performance.

1 Introduction

With the evolution of sensor technology, new devices and high performance communication networks, there is a clear need in managing these so-called 'smart things', as well as the vast amount of data being generated, in an efficient and transparent manner for the user, such that they can access services and gain insights that are relevant and actionable. This poses the question as to what are the new challenges brought about by this new demand in terms of engineering?

One way to address these new challenges is based on creating an abstraction layer, which is overlaying the physical infrastructure. This is often referred to in the literature as a *virtual sensor*, which has the role of isolating applications from the hardware by emulating the physical sensor in software [8]. In this manner, multiple logical instances of a physical sensor can exist for supporting various applications with different goals. So, one of the main utilizations is that of having a single sensor provide data to multiple applications and thus a large number of users with different

Malmö University, School of Technology, Sweden
Internet of Things and People Research Center
e-mail: {radu.c.mihailescu, jan.a.persson, paul.davidsson, ulrik.eklund} @mah.se

* Work partially supported by the Knowledge Foundation through the Internet of Things and People research profile.

© Springer International Publishing AG 2017   217
C. Badica et al. (eds.), *Intelligent Distributed Computing X*,
Studies in Computational Intelligence 678, DOI 10.1007/978-3-319-48829-5_21

objectives [6]. Another important aspect is to enable applications with a flexible way to provide services based on fusing sensor data either owned by the same provider or from multiple sensor infrastructure providers. Overall, virtual sensors facilitate concurrency in the sense of having different, possibly overlapping, subsets of sensors committed to different tasks.

A common example of deploying virtual sensors has to do with using sensed data from one location to then generalize it for other similar locations where sensors are not present. The same predictive approach can be used when certain sensors are malfunctioning. Such a solution is proposed in [3], where the authors present *VirtuS*, a prototype implementation of virtual sensor for TinyOS[2]. The system uses stacks to store data temporarily, offering the possibility to perform basic arithmetic operations (e.g. mean value of four data readings) before returning the value to the application. A scenario is presented to monitor the level of light inside a building.

Yet another frequent utilization of virtual sensors is that of removing the occurrence of outliers in readings, based on correlating data from a number of the same type of sensors, located in proximity of each other. As a result of this aggregation, not only are the readings more reliable but also, transmission is more efficient since less data would be transmitted onwards to the upper layers. This is particularly important in wireless sensor networks [5]. Similarly, in [2] a virtual sensor approach is employed to gathers track-data from several visual sensors and to determine a performance score for each sensor based on the coherence with the rest of the data. This information is feedbacked to the sensor, allowing it to use it as external information in order to reason about the confidence of its readings. Along the same lines, active noise control is an especially addressed topic in the context of virtual sensing. The challenge here is to create quiet zones around certain target points, without actually placing sensors at those desired virtual locations [7]. Among other applications, we also note the traffic domain, where virtual sensor are used to provide real-time congestion information, such as in the case of Washington state's traffic system [4].

The majority of the solutions in this space stand out in the sense that they provide a way to process homogeneous types of data, generally retrieved from a predetermined grouping of sensors. However, there are also a number of examples where virtual sensors combine heterogeneous data types for computing new values. Mobile devices, such as smartphones are a clear example of exploiting the sensors embedded in the device: accelerometers, gyroscope, GPS, microphone, camera, proximity sensors, ambient light sensors, and so forth to fuse the data and obtain an overall and more complete perspective of the environment and the users' activities [10]. In this work, we are not only concerned with combining data from different heterogeneous sensors in order to provide services to the user, but moreover, to design an extensible and reusable framework, that can determine at each point in time which are the most adequate data sources, given a certain description of the virtual sensor.

The organization of the rest of this paper is as follows. Section 2 introduces a new formalism for virtual sensors in terms of heterogeneous and distributed data

---

[2] TinyOS is an open-source operating system designed for wireless sensor networks

that is manipulated dynamically. We then provide in Section 3 the problem formulation based on the model introduced in the previous section. Also we provide a brief description of our implementation of the model, illustrated in a smart building scenario. In Section 4 we propose a framework that supports sensor collaboration for the task of efficiently allocating data sources to DIVS for improving the accuracy of measurement. Section 5 concludes.

2 The Dynamic Intelligent Virtual Sensor Model

In this section, we introduce a model of virtual sensors in response to the shortcomings identified in the previous section. In order to formalize our model, we present the following assumptions and definitions.

2.1 Definitions

The most basic type of sensor considered in our model is an individual hardware sensor that measures a single type of physical property in its environment.

Definition 1. *Individual Physical Sensors (IPS) are characterized by the following attributes* $\langle sid, frq, loc, pos, out \rangle$

- *sid - sensor ID*
- *frq - maximal sampling frequency*
- *loc - physical location being sensed*
- *pos - physical location of the sensor*
- *out - a tuple* $\langle t, p, u, r, d \rangle$ *describing the output*
    - *t - the type of physical property/quantity measured in the environment*
    - *p - precision of measurement*
    - *u - unit of measurement*
    - *r - range of values*
    - *d - data type*

An IPS is denoted in terms of a number of nominal values, which capture the sensor's capabilities, as well as other descriptive static information. It is important to point out the distinction between the attribute **loc**, which concerns the area it is monitoring (e.g. for a camera this means its associated field of view) and **pos**, which captures the current positioning of the sensor. Also, notice the distinction between the type of data being output, which determines the possible values for that type (e.g. numerical, categories, vectors, array of pixels), and the type of property in the environment being measured by the sensor.

Definition 2. *Dynamic Intelligent Virtual Sensors (DIVS) denote an abstract measurement resulted from combining a number of heterogeneous data sources provided by a group of physical sensors and/or other DIVS, whereas the set of sensors supporting a DIVS can change over time to better accommodate the current context and*

accuracy of the measurement. DIVS are characterized by the following attributes:
$\langle sid, I, M, L, C, O \rangle$

- *sid - sensor id*
- *I - a set of tuples $\langle p, w \rangle$ describing the input*
  - *p - set of relevant types of (physical) properties*
  - *w - information gain*
- *M - membership set of constituent sensors (sid)*
- *L - computational model $\langle P, E, Q \rangle$*
  - *P - set of preconditions*
  - *E - set of effects*
  - *Q - set of parameters of the model*
- *C - sensing context $\langle loc, TTL \rangle$*
  - *loc - localization of the monitored area*
  - *TTL - time to live of the DIVS*
- *O - a tuple $\langle t, AoM, u, f, r, d \rangle$ describing the output*
  - *t - the type of (physical) property measured in the environment*
  - *AoM - estimation of the accuracy of measurement*
  - *u - unit of measurement*
  - *f - sampling frequency*
  - *r - range of values*
  - *d - data type*

In contrast to IPS, the description of DIVS provides several extensions. Notice that a DIVS represents a software component defined recursively as a *dynamic* grouping of IPS and/or DIVS, which constitutes the membership set M, that is used to source *heterogeneous* types of properties according to the information in the input set I. The definition of the DIVS outlines the types of properties p which are considered in computing the output, in the set I. In other words, I takes values in the powerset of all possible IPS and DIVS outputs. For each type of property p specified in the input set I, we associate its *information gain* parameter, which represents the explanatory power of that particular property towards computing the output value. The information gain is assumed to be provided in the definition of the DIVS.

We can distinguish between attributes of DIVS that are determined at runtime, as opposed to those specified at design time. Specifically, the set M is instantiated at run-time, in terms of the identifiers (*sid*) of the sensors that are members of the DIVS and thus, providing an input stream of data. Also, the output set contains a dynamic part, namely the estimation of the accuracy of the measurement (*AoM*), which is an estimation of the current performance of the DIVS. The rest of the attributes are considered static. The DIVS applies a certain computation function, encapsulated by the preconditions-effects sets, in a rule-based manner, in order to generate the outputs. Context refers here to the spatio-temporal constraints that restrict the set of potential data sources to a certain subset based on the physical locations that they are currently monitoring, as well as their availability during the desired lifetime of the DIVS (*TTL*). Thus, the challenge becomes one of dynamically assigning data sources to DIVS (i.e. populating the membership set), such that at all times the accuracy of the measurement is maximized.

Example: To better illustrate DIVS, we give the following example of *climate monitoring* in a smart building environment. The input set specifies the required types of data $I = \{\langle temp(t), w_1 \rangle, \langle humidity(h), w_2 \rangle; \langle luminosity(l), w_3 \rangle\}$. Its precondition requires a certain precision of the temperature sensor and a certain sampling frequency of the luminosity sensor: $P = temp.prec \leq 0.1°C \wedge luminosity.freq \leq 1s$. The effect gives the formula for computing the climate *score* $E = compScore(t, h, l, w)$. The context sets the spatio-temporal constraints $C = \{12.00 - 13.00, Room09\}$.

Definition 3. *A DIVS system is defined as a tuple* $\langle S, D, Rep \rangle$*, where S represents a set of IPS, D represents a set of DIVS and Rep denotes a repository responsible for registering and providing information about all currently active IPS and DIVS.*

It is important to point out that a *DIVS system* requires that, on the one hand, multiple DIVS can subscribe to consume data from the same source, while on the other hand, multiple sensors (IPS or DIVS) can provide data to the same DIVS.

3 Problem Description

In the context of a DIVS system, we can formulate the problem as follows: given a DIVS description $\langle \textbf{sid, I, M, L, C, O} \rangle$, as well as a set of physical sensors S and a set D of DIVS, determine the membership set M of data sources $m \in S \cup D$, conditioned by the set of contextual constraints C, such that AoM is maximal. Solution updates are supposed to be generated on-the-fly in response to changes in the context data (e.g. mobile user changes room location, mobile sensors, sensor offline, etc.).

3.1 Inference Engine Analogy

Because in its simplest form, determining a solution essentially boils down to performing a search in a knowledge-base, comparing the problem to an inference engine is useful.

Suppose the knowledge-base KB contains information about all existing DIVS and IPS in the form of facts. A fact is a tuple that provides the descriptive attributes of a sensor. Let $fact1 = \{ips1; 1s; Room07; (temp, 0.05, °C, [-50, 50], float)\}$ denote a temperature sensor in a room. Now, lets consider the previous example in Section 2, having a DIVS for the task of monitoring the climate conditions. Then, obtaining the membership set can be posed as a query $Q = \{?temp, ?humidity, ?luminosity\}$ in KB, subject to the set of preconditions. This means, find the tuples whose output type matches the input type of the DIVS, via binding the variable $?var$ to a value in KB. If the knowledge-base contains only $fact1$, the query will return $ans = \{ips1, ?humidity, ?luminosity\}$. Binding a variable in the membership set ensures that the output of that sensor is transmitted as input data to the DIVS. The same principle can be applied to perform more complex inferences such in the case of chaining a number of sensors outputs in order to satisfy the input requirements of a DIVS.

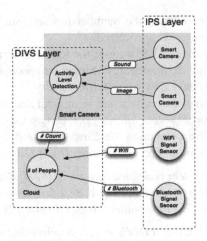

Fig. 1: An example of an implemented DIVS structure and data flow.

With this analogy, the problem of selecting an instantiation of the membership set translates into a query in a knowledge-base in a straightforward manner. However, in this work we consider a more general scenario, where we do not assume that all of the required types of data are always readily available. This brings about two important aspects: *(i)* when a particular input data type provided in the DIVS specification is not available, but may be approximated using readings from other types of sensors and *(ii)* when new data types are identified, based on which the accuracy of the measurement can be improved. In other words, the question is how can DIVS support sensor collaboration for accomplishing more complicated tasks.

3.2 A DIVS Application to Smart Buildings

In this section we illustrate an application of our proposed DIVS model in the context of activity monitoring using sensors in a smart building environment. Figure 1 depicts the overall structure of the developed system. The goal here is to estimate the usage of a certain area of the building in terms of the number of people. The business value of providing this information in the application can range from evaluating the utilization of conference rooms, to scheduling cleaning teams, to adapting the HVAC (heating, ventilation, and air conditioning) control of the building for increased energy efficiency. In this sense, DIVS can be designed on-the-fly to exploit the existing sensing infrastructure in order to support new value-added services.

Specifically, the system consists of an IPS that provides a count of the number of wifi-enabled devices in the area and another IPS that is responsible for monitoring the number of activated bluetooth devices. Moreover, we deploy a smart-camera that implements one IPS outputting the sound level for the designated area and another IPS which performs an image-based person count for the camera's field of view (FOV). Now, the previous two IPS outputs, representing heterogeneous types of data, are aggregated by a DIVS, which carries out the task of implementing an activity level detection to estimate the overall number of people in the area. The DIVS computation includes an image preprocessing phase of determining the movement directions of a person and computing the difference between the incoming and outgoing events with respect to the specified perimeter.

Based on the enumerated components, we describe the main DIVS, which captures the purpose of our application. For the sake of simplicity, the definition of the main DIVS specifies that the output, representing the estimated number of people in real-time, is computed as a linear combination of the input variables[3]. Essentially, in our implementation, the data is stored to a cloud service and a REST application consumes queued data and generates estimations which are then persisted in the cloud. This does not require though that DIVS are always implemented in the cloud. For instance, as seen in Figure 1, in our scenario, the smart-camera is running a DIVS which implements the activity level detection using both image and sound as inputs. However, the system must be able to integrate all of these components.

Now, suppose for instance that a number of additional sensors are available on the premise, monitoring temperature, humidity, light intensity, carbon dioxide particles, air pressure and movement. According to our problem description, the challenge is then to determine whether the accuracy of the output can be improved by using some of the available sensor types that have not been explicitly prescribed as inputs. Alternatively, if one of the inputs in the definition of the DIVS is not available, can we substitute it using one or more sensor types? Moreover, what is the information gain that could be associated to these newly acquired inputs, for further processing during the data fusion phase? Recall that a DIVS can also take as input the output of another DIVS. For example, a climate monitoring DIVS, similar to the example in Section 2, that polls data from a number of the abovementioned sensors can provide useful information with regard to the reusability of some of its inputs or its outputs. To address these questions we propose in the next section a framework to solve the dynamic sensor allocation problem based on a semantics-empowered search heuristic.

4 Collaborative Sensing Framework

In this section we put forth a framework designed to identify the best allocation of sensor inputs for obtaining a maximal *AoM* for a given DIVS. This functionality is encapsulated by the DIVS controller. Figure 2 shows the architecture of the proposed framework. The main components of the architecture are presented hereafter.

As explained in Section 3.1, the *Inference engine* has the role of acting as a template selecting the sensors in set M that match the predefined types specified in the input set I. In addition, sensors are selected in compliance with the context conditions. Namely, the monitored area of the DIVS includes the subareas of its constituent sensors. Thus, the membership set M is updated whenever the context of the DIVS has changed (e.g. a DIVS running on a smartphone changing location),

---

[3] Clearly, more complex data fusion strategies could be considered to this end.

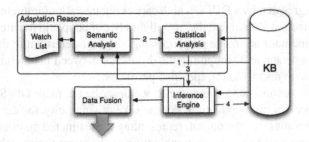

Fig. 2: DIVS Controller for generating and updating the membership set M (the numbering corresponds to the phases in the CBR cycle)

or the constituent sensors themselves have modified their *loc* parameter (i.e. mobile sensor). This information is retrieved by the *Inference engine* from the *KB*, which collects and provides these details in real-time.

Definition 4. *A solution is **complete** if all types of properties in the set I are associated to a certain data source. Otherwise, the solution is considered **incomplete**.*

If the solution is complete, then the input streams are fed directly to the *Data fusion* block. We do not dwell in particular in this work on the data fusion techniques since they are generally highly specific to the application domain. Thus, we assume this component is specified in the definition of the DIVS. However, even though the solution may be complete, the *Adaptation Reasoner* block carries out an open-ended procedure of attempting to improve the *AoM* metric. Also, it has the role of computing a solution when the current one is incomplete, meaning that not all of the data types in I are available.

4.1 Case-Based Reasoning

To address the goals of the *Adaptation Reasoner* we propose a case-based reasoning (CBR) approach, whereas a new problem can be solved by searching in a knowledge-base for similar precedents and adapting their solutions to fit the current problem. In order to compute a solution we apply the following CBR cycle: *(1)* retrieve most similar cases using semantic matchmaking; *(2)* reuse information from those cases to compute a solution; *(3)* revise this information according to the current case; *(4)* retain new information by incorporating it into the knowledge-base.

The *Adaptation Reasoner* includes on the one hand a *Semantic analysis* component and on the other hand a *Statistical analysis* component. In terms of the former component, as *semantics* we understand the introduction of machine interpretable languages in the descriptions of resources, by the use of ontologies and semantic markup language, which provide a formal and explicit specification of shared concepts. We require henceforth that the description of DIVS is augmented with semantic information such that the sets, I,O,P,E consist of semantic concepts based on a shared primitive term vocabulary V. This has the role of enabling automatic discovery and composability of DIVS in a straightforward manner. The degree of match between two semantic sets is computed via a bipartite matching graph [1].

Let two DIVS $\mathscr{D}_i$ and $\mathscr{D}_j$ with two respective sets of concepts C_i and C_j and a complete, weighted bipartite graph $G = (C_i, C_j, E)$ that links each concept $c_i \in C_i$ with each concept $c_j \in C_j$, $(c_i, c_j) \in E$, where $E = C_i \times C_j$ represents the edges in the graph. The weight ω_{ij} associated to each (c_i, c_j) edge represents the semantic similarity between those concepts. Four degrees of match are generally considered: *exact, subsumes, plug-in, and fail* [9]. The match takes the value *exact* if the concepts c_1 and c_2 are equivalent $c_1 \equiv c_2$; *subsumes* if c_1 subsumes c_2, $c_1 \sqsupseteq c_2$; *plug-in* if c_1 is subsumed by c_2, $c_1 \sqsubseteq c_2$; *fail*, otherwise. Each degree of match is assigned a value in the interval $[0, 1]$, where 1 represents an exact match among the terms, 0.75 represents a subsumes relation, 0.5 represents a plug-in relation, and 0 a fail. The maximum weighted bipartite matching $G\prime = (C_i, C_j, E\prime)$ is obtained by retaining the edges (best match among concepts) that have the maximal value. The graph $G\prime$ is called a relaxed bipartite graph because not all the concepts from C_i have to be connected to a concept in C_j. Then, the degree of match is the weight of this graph, which is calculated with the following formula:

$$W_{G\prime} = \frac{\sum_{\omega_{ij} \in E\prime} \omega_{ij}}{max(|C_i|, |C_j|)}$$

The degree of match between two DIVS can be regarded as an indication of the similarity and thus, the *information gain* that one can provide in determining the output of the other. Thus, stage *(1)* in the CBR cycle is concerned with performing a semantic matchmaking on the description of DIVS in order to retrieve the most similar cases. If the degree of match exceeds the 0.75 threshold, meaning that $\mathscr{D}_i$ subsumes $\mathscr{D}_j$, then the output of $\mathscr{D}_j$ is used directly as input for computing the output of $\mathscr{D}_i$, by assigning it an information gain equal to their degree of match.

Otherwise, in case $\mathscr{D}_i$ is subsumed by $\mathscr{D}_j$, this means that the semantic relation between the two DIVS entails that the concepts of $\mathscr{D}_j$ include those in $\mathscr{D}_i$. Hence, $\mathscr{D}_i$ may potentially exploit some of the inputs of $\mathscr{D}_j$. In phase *(2)*, the *Statistical analysis* block evaluates whether to include into the membership set M some of the input data types considered relevant by semantically similar DIVS. This is achieved by computing the *Pearson* coefficient, which gives a measure of the correlation between two variables. In case the result shows a certain correlation, but is not satisfactory, the input is appended to the *Watch list*, which holds a record of promising inputs. These candidates are further inspected periodically. The *Pearson* coefficient represents then the information gain for that particular input. However, if the match is a *fail*, the inputs and outputs of $\mathscr{D}_j$ are disregarded in the computation of $\mathscr{D}_i$. The *Statistical analysis* block is also responsible for providing the *AoM* estimation.

During the third stage *(3)*, once the inputs, alongside their associated information gains are determined, the DIVS applies a computation model, encapsulated by the preconditions-effects, in a rule-based manner, in order to generate the outputs. Notice that different data fusion techniques can be prescribed at this stage, which could even be executed alternatively based on their associated preconditions. A straightforward manner of using the input data would be to perform a linear combination, where the weights are represented by the information gains of each input. Finally, in stage *(4)* the solution obtained is stored to the knowledge-base and made available for further queries.

5 Conclusions

Summarizing, this work puts forward a novel approach to virtual sensors, which takes into consideration the aggregation of heterogeneous types of data, as well as the dynamic aspect of determining at each point in time the best allocation of sensors for a particular task with respect to the optimization of the accuracy of the output data. This paper shows how such systems can be engineered according to the model of DIVS to provide a wide range of complex services. In particular, we present a system developed for the task of assessing the number of people for a certain area, in a smart building context. We then extend the model by emphasizing two instances when the system may operate suboptimal given the underlying sensor infrastructure. Furthermore, we propose a versatile framework that employs semantics to deal with of lack of input data types and low accuracy of the output of the DIVS.

References

1. U. Bellur and R. Kulkarni. Improved matchmaking algorithm for semantic web services based on bipartite graph matching. In *Proceedings of the IEEE International Conference on Web Services (ICWS)*, pages 86–93, 2007.
2. F. Castanedo, J. Garcia, M.A. Patricio, and J.M. Molina. A multi-agent architecture to support active fusion in a visual sensor network. In *Proceedings of the Second ACM/IEEE International Conference on Distributed Smart Cameras (ICDSC)*, pages 1–8, Sept 2008.
3. P. Corsini, P. Masci, and A. Vecchio. Configuration and tuning of sensor network applications through virtual sensors. In *Proceedings of the Fourth Annual IEEE International Conference on Pervasive Computing and Communications Workshops (PerCom)*, pages 315–320, 2006.
4. Daniel J. Dailey and Frederick W. Cathey. Deployment of a virtual sensor system, based on transit probes in an operational traffic management system. Technical report, Washington State Transportation Center, 2006.
5. Md. Motaharul Islam, Mohammad Mehedi Hassan, Ga-Won Lee, and Eui-Nam Huh. A survey on virtualization of wireless sensor networks. *Sensors*, 12(2):2175–2207, 2012.
6. Ilias Leontiadis, Christos Efstratiou, Cecilia Mascolo, and Jon Crowcroft. Senshare: Transforming sensor networks into multi-application sensing infrastructures. In *Proceeding of the 9th European Conference on Wireless Sensor Networks (EWSN)*, pages 65–81. Springer Berlin Heidelberg, 2012.
7. Lichuan Liu, S. M. Kuo, and M. Zhou. Virtual sensing techniques and their applications. In *Proceedings of the International Conference on Networking, Sensing and Control, (ICNSC)*, pages 31–36, 2009.
8. S. Madria, V. Kumar, and R. Dalvi. Sensor cloud: A cloud of virtual sensors. *IEEE Software*, 31(2):70–77, 2014.
9. Massimo Paolucci, Takahiro Kawamura, Terry R. Payne, and Katia Sycara. Semantic matching of web services capabilities. In *Proceedings of the First International Semantic Web Conference (ISWC)*, pages 333–347. Springer Berlin Heidelberg, 2002.
10. Z. Yan, V. Subbaraju, D. Chakraborty, A. Misra, and K. Aberer. Energy-efficient continuous activity recognition on mobile phones: An activity-adaptive approach. In *Proceedings of the 16th International Symposium on Wearable Computers (ISWC)*, pages 17–24, 2012.

Intelligent Data Metrics for Urban Driving with Data Fusion and Distributed Machine Learning

Fábio Silva[1], Artur Quintas[1], Jason J. Jung[2], Paulo Novais[1], and Cesar Analide[1]

[1] Algoritmi Centre, University of Minho, Braga, Portugal
fabiosilva@di.uminho.pt, arturffq@gmail.com, pjon@di.uminho.pt,
analide@di.uminho.pt
[2] Department of Computer Engineering, Chung-Ang University, Seoul, Korea
j2jung@gmail.com

Abstract. Using a community of users allows the collection of data that can be used towards the benefit of society. Aligned with trends such as smart city design and the internet of things, the range of application are being only restricted by human imagination. Taking the case of urban driving, it is already possible to estimate roadblocks, congestions and issue real-time alerts to users using popular applications. The approach taken in this paper, furthers this analysis by providing means to analyse the root cause of not only such events but also dangerous driving habits from users. Making use of machine learning algorithms, big data and distributed systems, a work-flow based on the PHESS Driving platform was developed. Results achieved are satisfactory in the field tests produced, giving reason to some popular common sense, as well as, new theories for dangerous driving events.

Keywords: Intelligent Systems, Urban Driving, Smart Cities, Ubiquitious Computing

1 Introduction

Computer platforms exist to ease hard labour tasks from human life. From its inception in mathematical calculus, these platforms have evolved taking advantage of advances on hardware and software. In current times, concepts such as smart cites and the internet of things hide complex and intricate computer networks to ease human life.

Perhaps, one is the increasing presence of sensors in our physical world, both integrated into existent infrastructure or mobile devices. In this approach, this work offers an alternative to gather information about urban driving with the intent to provide users and communities with possible explanations to dangerous driving events. For such, prior work on driving assessment is used to obtain driving data and event classification is used as an input in a data fusion project to create machine learning models and pattern recognition algorithms [18] [19].

1.1 Smart City Design

There are mainly two trends in how Smart Cities are defined, the focus on a single urban aspect, such as an ecological approach, and the focus on the integration of various interconnected urban aspects [16]. Broadly speaking, a Smart City refers to a city which makes use of Information and Communication Technologies (ICT) to enable urban improvement.

An example of the first trend is the Smart Grid project in Málaga, Spain, which makes an ecological approach to the Smart City. In 2014, the project was able to reach its goals of 20% energy savings, and a reduction of CO_2 emissions of 6,000-tonne per year [8].

One of the largest smart city experimental testbeds is the city of Santander in Spain. The project, named SmartSantander, deployed over thousands of sensors around the area of Santander so as to monitor the real-time state of the city. These sensors measure many environmental parameters such as light, temperature, and noise, as well as parameters like the occupancy of some parking slots[20].

In the European Santander and Genova cities along with the Japanese cities of Mitaka and Fujisawa, the ClouT (*Cloud of Things for empowering the citizen clout in smart cities*) [9][22] project is an example of a citizen-centric multi aspect approach to a Smart City. It attempts to classify use cases of one of the three types found to be relevant after discussing with stakeholders, those types being: Smart city resource management, Safety and emergency management, and Citizen health and comfort management [24].

Using data from a lot of sensors can demand analytics to be performed over Big Data. In this context, Big Data refers to data sets that are so large or complex that they demand an ability of data processing and analysis that goes beyond the typical database tools [15]. A platform named City Data and Analytics Platform (CiDAP) [4] attempts to deal with this issue in the context of Smart Cities. This platform stores, processes, and analyses the data generated by the aforementioned SmartSantander project, focusing on dealing both with historical data and real time data.

Another interesting project that makes use of the Santander testbed is a project that attempts to answer the question: "Are the data from the Smart Santander consistent enough to predict a traffic jam and build a project to improve traffic management in the city?"[23]. The authors are able to point out traffic jams with 99.95% accuracy using multiple algorithms combining three prediction methods, namely using the Tree Ensemble, Fuzzy Rule, and Probabilistic Neural Network. It is of relevance to note that, the most important historical feature identified to predict the traffic status, is the existence of rain.

1.2 Urban Driving Approaches

There are different trends to assessing driving and traffic flow in the literature. Tango and Botta [21] identify driver's distraction, in a simulated environment, using speed, time to collision, steering angle, lateral position, position of the accelerator pedal, and position of the brake pedal. Jiao et al.[12] looks for signs of driver fatigue by analysing the Slow Eye Movement (SEM) using an electro-oculogram in combination with a driving simulator. Yoshida et al. [25] attempts to characterize a driver's cognitive state in real driving situations using the vehicle's Controlled Area Network (CAN) and an eye movement tracking device in order to predict the steering entropy and task cognition. In Jabon et al.[11], using a video feed in a driving simulator the authors attempt to predict unsafe driving behaviour by analysing facial expression of the driver in combination with the simulation information regarding aspects such as road conditions, acceleration and due to pedal presses and braking, and type of accident.

Another trend is to use mobile devices as driver sensing platforms. Regarding the driver behaviour the MIROAD [13] is a system that uses various inputs from a smartphone, like GPS, accelerometer, camera, and gyroscope. These inputs are used to detect events and discern their nature as being either typical or aggressive. Castignani et al.[3] attempts to generate a single score for both driving inefficiency and aggressiveness. The sensors used were the GPS, the accelerometer, the magnetometer, and the gravity sensor. The results presented suggest that in urban areas drivers tend to drive aggressively with higher overspeed and acceleration events.

The PHESS Driving platform [18] [19], uses kinetic indicators relating to the amount of accelerations and brakes, velocity, and intensity of acceleration and breaking as well as the force exerted in the vehicle during curves to discern driving style between three categories ordered by magnitude. Based on the driving profile of each person, namely the gravity of each driving event, a community profile is then built relating to squared map regions that form a grid.

Evaluation through regions is also adopted by Dolui et al.[6], in this case the authors attempt to determine the status of traffic of a particular region using real time data concerning GPS points and calculating the current speeds from the positions and timestamps. Making use of the accelerometer to find out if slowing was due to a pothole or a speed breaker, the authors categorise traffic as Open, Normal, Slow, Congested, or Impending. Akhtar et al.[1] opted for fuzzy inference for evaluation of data. Using fixed axes, the authors derive information regarding the magnitude of acceleration while changing lanes (X axis), vehicle acceleration and deceleration (Y axis), and road anomalies(Z axis). Saiprasert and Pattara-Atikom [17] make a GPS based approach to detect speeding. The elements retrieved from the recorded GPS data are a measure of instantaneous speed of the vehicle at each data sample, the position composed by the latitude and longitude of each data sample and the heading which is the direction in degrees East of True North. The authors go on to compare speed recorded in the car's speedometer and calculated from the smartphone's GPS, in their findings there seems to be a constant offset of speed thus suggesting a similar accuracy.

LuxTraffic [14] is a Luxembourg based project that tries to convey real time traffic information using smartphone sensor data. A static approach is made where mapped road checkpoints are used together with the smartphone GPS and location services to determine in which road segment a vehicle is in. Calculating the speed using distance over time, an aggregation of the speeds in a road segment can then characterize the traffic flow of a road segment. The validity of input data is evaluated depending on factors such as being rush hour or weekend. The platform uses the external service [5] to complement its information when there are segments with absent data.

Related with traffic conditions and driving performance, Gonçalves et al.[10] make use of the speed limit for two sorts of indicators, first a driver's performance is evaluated depending on whether or not the driver goes over the speed limit, and secondly, if in a time window the average of the vehicles is equal or above the speed limit meanings low traffic congestion, and that the lower the vehicle's velocity the higher traffic congestion. Road conditions is also a subject of interest related to road traffic expression, Astarita et al.[2] model the state of probable road anomaly depending on the variation of vertical acceleration and sound. Based on experimental results a lower bound for acceleration and upper bounds for both acceleration and sound are established. Still related to road conditions, Douangphachanh and Oneyama[7] attempt to estimate the International Roughness Index(IRI) using a smartphone and Vehicular Inteligent Monituring System (VIMS) to establish the actual IRI for comparison in their analysis.

2   Intelligent Data Metrics for Urban Driving

The implementation of an intelligent platform to influence driving behaviour encompasses different approaches. From psychological strategies such as rewarding positive behaviours and punishing negative behaviours to leading by example. The PHESS Driving platform [18] tries to tackle the problem of urban driving with different modules. The intelligent data metrics module is one specialised in increasing not only the quality of information but also its influence over the driver. An illustration of this module can be found in Figure 1.

Integrated in a distributed platform, it allows the acquisition, processing and storage of data in multiple devices with graceful degradation of services, meaning if one component ceases to work the rest of platform can still operate partially without the need to completely stop. The architecture presented is based on such principles. The point of failure is based on the number of computing and sensing devices.

To obtain precise data, a methodological approach to data treatment was needed and evidenced in the architecture. This leads to the idea of a processing pipeline, like machine learning work-flows, it entails a series of steps towards the preparation and fusion of data and creation and update of machine learning models.

The purpose of the Intelligible Data Metrics (IDM) module is to look at road traffic data to both enrich trip information and perform analysis that is able to model the expectation of problematic driving behaviour based on the context surrounding an event instead of the vehicle conditions or driver state.

This module is divided in two components, the data aggregation and data analysis. The first component enriches a PHESS Driving trip with contextual data. This data consists of local date and time, weather information, GPS points, and neighbouring locations of interest. New data is obtained by making requests to services like the PHESS Driving API, Google Maps Timezone API, the Weather Underground historic API, the Track Matching API, and the OpenStreetMap API.

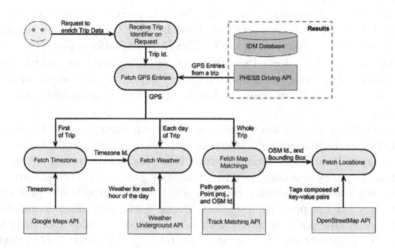

Fig. 1. Data flow in the data aggregation component.

The aggregation is pivoted around the GPS entry, which contains the latitude and longitude coordinates and the recorded time. The date time information concerns minute, hour, day of week, day of month, month, and year. It is obtained by discovering the timezone identifier of the whole trip, and then constructing the date time for each GPS entry. Weather information contains the following aspects: temperature, wind speed, wind direction, humidity, and precipitation.

The location elements retrieved from the OpenStreetMap (OSM) server are cached as a grid of regions in the database. For each GPS entry, an area surrounding the road segment where that the entry belongs to, is computed. This is then queried in the database for intersection with location elements originated from the OSM.

On the data analysis component, the aggregated data is fused into a singular representation of an entity in the form of a feature vector. The numeric values are stored without change on this feature vector. The textual values, namely the properties of the locations of interest retrieved from the OSM, are mapped into feature indexes and take binary values of 1 and 0 respectively representing the presence or absence of the property in an entry.

3   Evaluation of Machine Learning Models

The evaluation of machine learning models is essential to determine their usefulness. For instance, if the aim is to predict bad driving events, then a higher cost for not predicting a bad behaviour (false negative) than predicting a bad event erroneously (false positive) should be used. The approach in this research, follows different evaluation strategies to demonstrate and explain the usefulness of the platform developed.

The solution is envisioned as a classification problem where every entry is classified as either problematic or not. This classification is based on the magnitude of trip events according to the PHESS Driving platform [18].

The point of view adopted was that of a city attempting to evaluate several scenarios and allocating resources to solve any expected problematic situations. The focus was to find out issues and to be sure that these problems are worth checking.

The evaluation metrics compared were precision, accuracy, and recall. Precision indicates the likeliness of a true positive, this is essential to the task at hand since it is costly to spend time and effort fixing something that is not in fact a problem. Accuracy was computed as a means of giving an overall picture of the model's performance, but since the data set is skewed towards non problematic events, it's not the preferred indicator. Recall is the percentage between true positives and true positives added with false negatives. It allows for there to be a context to the precision, it indicates how many problems are being caught of the whole set of problems.

After modelling the data in relation to date time, weather, and neighbouring location with the goal of identifying problematic driving situations, this model is used by the IDM system to evaluate new entries. The evaluation of each entry can produce a report that indicates the combination of factors that lead to the attributed classification. An example of this report is showcased in a demonstration web page that provided, GPS, date time, and weather information on a form complements it with neighbouring locations on the server side and produces a report like the one shown in Figure 2.

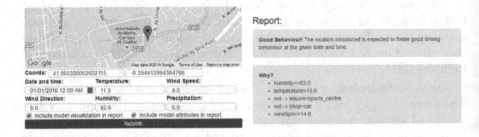

Fig. 2. Report shown in the demonstration web page.

Experiments detailed in this section used the PHESS Driving user records from 2014 and 2015, entailing a total of 7 drivers without restriction on their driving activities. This means, there was no restriction on daily and night driving, on congested and not-congested routes or any other means. A total of 70 trips were recorded for research proposes during this period, which produced a total of 200 kilometres of driving experience over the city of Braga located in Portugal. A different set of smartphones were also used to log driving trips, composed of 8 different smartphones from different brands and models.

Model generation used a server machine equipped with intel i7 processor and 8GB of memory to produce machine learning models. As a machine learning library (MLLib), Apache Spark was used due its compliance with distributed computing philosophy and the ability to produce meaningful results.

Four scenarios were contemplated, the main use case of the IDM system which uses features belonging to all three aspects, date time, weather, and neighbouring locations is the first scenario. So as to evaluate the predictive power of each of those three separate sets of features three more scenarios were evaluated each using only the features corresponding to its set.

In order to evaluate the results, 30% of the whole dataset was held out of the training phase and then used to test the trained models.

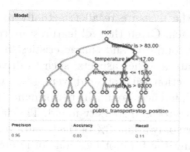

Fig. 3. Example of a trained decision tree.

Figure 3, exemplifies a trained decision tree, which uses context data to generate a green or red classification about the context upon which the driving event is being performed. The nodes represent contexts and the route from the root node to the leaf node represents a context condition with a classification on the leaf node.

To contextualize the results a few other models were used. As a baseline the coin toss model was used, where each entry is randomly decided to be problematic or not. And the all in and all out models which respectively refer to considering all entries as problematic and no entries as problematic. Besides the aforementioned coin toss approach two more were made so as to provide comparison of what is the tradeof, if any, of using the Decision Tree for its intelligible nature. These approaches were training and using a Naive Bayes model, and a linear Support Vector Machine (SVM) model. Furthermore the algorithm Decision Tree has been trained using two different strategies for the measure of the

homogeneity of the labels at each node: gini impurity and entropy available at
the Apache Spark machine learning tool.

Method	Precision	Recall	Accuracy
Coin Toss	0.166	0.511	0.484
DT(gini)	0.977	0.182	0.861
DT(entropy)	0.991	0.133	0.853
Naive Bayes	0.222	0.519	0.612
SVM	0.275	0.168	0.785

Table 1. Results using all three sets of features.

As can be seen in Table 1, the Decision Tree tends to perform very well in
comparison to the other tested methods. As results are heavily skewed towards
negative entries, i.e. entries which are not considered problematic, there is a still
an improvement of accuracy over the naive approaches. In most runs there was
no noticeable difference between the Decision Tree created using Gini or Entropy
as impurity measures. The precision obtained was considerably larger than the
baseline approaches, but the recall was higher using the All Positive, Coin Toss,
and Naive Bayes model.

A small value for the measure recall is actually to be expected given the
restraints of this approach. Given that all features are contextual, and none relates to the driver, his condition and characteristics. It is expected that many
problems which originate in other aspects, such as driver fatigue or driver experience, or even interaction with other drivers, seem to be beyond the scope
of this solution. The differences found in performance may also be related to
the size of the feature vector and sparsity found. The dataset tended to be very
sparse, where on average only about 2% of the feature vector had explicit values.
This is because encoding the type of locations of interest resulted in over 900
flags, each a feature.

Running tests separately with each dataset lead to the conclusion that on its
own the neighbouring location data has a very small recall for all models, the
highest being the Naive Bayes model with 7.3% with the Decision Tree model
just under 1% and the SVM incapable of discerning any Positive value. The date
time and weather data on their own had slightly worse results than those reached
with all three sets of features for the Decision Tree, which suggests there is a
benefit to using their composition as opposed to focusing on a single aspect.

4   Conclusion

The study present in this paper demonstrates the application of external contexts acquired from data fusion processes to predict and explain driving events.
This analysis is particularly aimed towards the analysis of user and community
data in order to find justification both individual to the user and general to the
community. The data used is derived from field experiments upon the PHESS

Driving platform and results reinforce the idea of the influence of other contexts aside from driver and vehicle to justify hazardous driving events. From the analysis presented, it possible to deduce that the models produced using machine learning are able to explain in a meaningful significance the expected behaviour of drivers using attributes such as time, weather and location.

The implementation of this system allow for the prevention of dangerous driving behaviour while using the mobile applications such as the PHESS Driving. As future improvements there are the consideration of more external contexts and the enhancement of the predictive mechanism by projecting the routine trips of users and cataloguing segments of the trip according to dangerous possibility and generating safer alternatives. Ensuring the continuous learning and estimating the validity of the generated models is also a point of concern in future analysis.

Acknowledgements

This work has been supported by FCT – Fundação para a Ciência e Tecnologia within the Project Scope UID/CEC/00319/2013

References

1. Akhtar, N., Pandey, K., Gupta, S.: Mobile application for safe driving. In: Communication Systems and Network Technologies (CSNT), 2014 Fourth International Conference on. pp. 212–216. IEEE (2014)
2. Astarita, V., Festa, D.C., Mongelli, D.W.E., Tassitani, A.: New methodology for the identification of road surface anomalies. In: Service Operations and Logistics, and Informatics (SOLI), 2014 IEEE International Conference on. pp. 149–154. IEEE (2014)
3. Castignani, G., Frank, R., Engel, T.: An evaluation study of driver profiling fuzzy algorithms using smartphones. In: Network Protocols (ICNP), 2013 21st IEEE International Conference on. pp. 1–6. IEEE (2013)
4. Cheng, B., Longo, S., Cirillo, F., Bauer, M., Kovacs, E.: Building a Big Data Platform for Smart Cities: Experience and Lessons from Santander. In: 2015 IEEE International Congress on Big Data (BigData Congress). pp. 592–599 (Jun 2015)
5. Cita.lu: Cita.lu | motorway network (2016), http://www.cita.lu/en
6. Dolui, K., Mukherjee, S., Datta, S.K.: Traffic status monitoring using smart devices. In: Intelligent Interactive Systems and Assistive Technologies (IISAT), 2013 International Conference on. pp. 8–14. IEEE (2013)
7. Douangphachanh, V., Oneyama, H.: Estimation of road roughness condition from smartphones under realistic settings. In: ITS Telecommunications (ITST), 2013 13th International Conference on. pp. 433–439. IEEE (2013)
8. Endesa: Smartcity Malaga - A model of sustainable energy management for cities of the future (2014), http://www.endesasmartgrids.com/images/pdfs/2014-02-14-white-paper-sm-malaga.pdf
9. Galache, J., Yonezawa, T., Gurgen, L., Pavia, D., Grella, M., Maeomichi, H.: ClouT: Leveraging Cloud Computing Techniques for Improving Management of Massive IoT Data. In: 2014 IEEE 7th International Conference on Service-Oriented Computing and Applications (SOCA). pp. 324–327 (Nov 2014)

10. Goncalves, J., Goncalves, J.S., Rossetti, R.J., Olaverri-Monreal, C.: Smartphone sensor platform to study traffic conditions and assess driving performance. In: Intelligent Transportation Systems (ITSC), 2014 IEEE 17th International Conference on. pp. 2596–2601. IEEE (2014)

11. Jabon, M.E., Bailenson, J.N., Pontikakis, E., Takayama, L., Nass, C.: Facial expression analysis for predicting unsafe driving behavior. IEEE Pervasive Computing (4), 84–95 (2010)

12. Jiao, Y., Peng, Y., Lu, B.L., Chen, X., Chen, S., Wang, C.: Recognizing slow eye movement for driver fatigue detection with machine learning approach. In: Neural Networks (IJCNN), 2014 International Joint Conference on. pp. 4035–4041. IEEE (2014)

13. Johnson, D., Trivedi, M.M., et al.: Driving style recognition using a smartphone as a sensor platform. In: Intelligent Transportation Systems (ITSC), 2011 14th International IEEE Conference on. pp. 1609–1615. IEEE (2011)

14. Kovacheva, A., Frank, R., Engel, T.: Luxtraffic: A collaborative traffic sensing system. In: Local & Metropolitan Area Networks (LANMAN), 2013 19th IEEE Workshop on. pp. 1–6. IEEE (2013)

15. Manyika, J., Chui, M., Brown, B., Bughin, J., Dobbs, R., Roxburgh, C., Hung, A., Brown, B.: Big data: The next frontier for innovation, competition, and productivity (2011)

16. Monzon, A.: Smart cities concept and challenges: Bases for the assessment of smart city projects. In: Smart Cities and Green ICT Systems (SMARTGREENS), 2015 International Conference on. pp. 1–11 (May 2015)

17. Saiprasert, C., Pattara-Atikom, W.: Smartphone enabled dangerous driving report system. In: System Sciences (HICSS), 2013 46th Hawaii International Conference on. pp. 1231–1237. IEEE (2013)

18. Silva, F., Analide, C., Novais, P.: Assessing road traffic expression. International Journal of Artificial Intelligence and Interactive Multimedia 3(1), 20–27 (2014)

19. Silva, F., Analide, C., Novais, P.: Traffic Expression through Ubiquitous and Pervasive Sensorization - Smart Cities and Assessment of Driving Behaviour. In: 5th International Conference on Pervasive and Embedded Computing and Communication Systems. pp. 33–42 (feb 2015)

20. SmartSantander.eu: SmartSantander, http://www.smartsantander.eu/

21. Tango, F., Botta, M.: Real-time detection system of driver distraction using machine learning. Intelligent Transportation Systems, IEEE Transactions on 14(2), 894–905 (2013)

22. Tei, K., Gurgen, L.: ClouT : Cloud of things for empowering the citizen clout in smart cities. In: 2014 IEEE World Forum on Internet of Things (WF-IoT). pp. 369–370 (Mar 2014)

23. Treboux, J., Jara, A., Dufour, L., Genoud, D.: A predictive data-driven model for traffic-jams forecasting in smart santader city-scale testbed. In: 2015 IEEE Wireless Communications and Networking Conference Workshops (WCNCW). pp. 64–68 (Mar 2015)

24. Yonezawa, T., Galache, J., Gurgen, L., Matranga, I., Maeomichi, H., Shibuya, T.: A citizen-centric approach towards global-scale smart city platform. In: 2015 International Conference on Recent Advances in Internet of Things (RIoT). pp. 1–6 (Apr 2015)

25. Yoshida, Y., Ohwada, H., Mizoguchi, F., Iwasaki, H.: Classifying cognitive load and driving situation with machine learning. International Journal of Machine Learning and Computing 4(3), 210–215 (2014)

A Probabilistic Sample Matchmaking Strategy for Imbalanced Data Streams with Concept Drift

Jesus L. Lobo[1], Javier Del Ser[1,2], Miren Nekane Bilbao[2],
Ibai Laña[1], and S. Salcedo-Sanz[3]

[1] TECNALIA. OPTIMA Unit, E-48160 Derio, Spain
{jesus.lopez,javier.delser,ibai.lana}@tecnalia.com
[2] University of the Basque Country UPV/EHU, 48013 Bilbao, Spain
{javier.delser,nekane.bilbao}@ehu.eus
[3] University of Alcala, 28801 Madrid, Spain
sancho.salcedo@uah.es

Abstract. In the last decade the interest in adaptive models for non-stationary environments has gained momentum within the research community due to an increasing number of application scenarios generating non-stationary data streams. In this context the literature has been specially rich in terms of ensemble techniques, which in their majority have focused on taking advantage of past information in the form of already trained predictive models and other alternatives alike. This manuscript elaborates on a rather different approach, which hinges on extracting the essential predictive information of past trained models and determining therefrom the best candidates (intelligent sample matchmaking) for training the predictive model of the current data batch. This novel perspective is of inherent utility for data streams characterized by short-length unbalanced data batches, situation where the so-called trade-off between plasticity and stability must be carefully met. The approach is evaluated on a synthetic data set that simulates a non-stationary environment with recurrently changing concept drift. The proposed approach is shown to perform competitively when adapting to a sudden and recurrent change with respect to the state of the art, but without storing all the past trained models and by lessening its computational complexity in terms of model evaluations. These promising results motivate future research aimed at validating the proposed strategy on other scenarios under concept drift, such as those characterized by semi-supervised data streams.

Keywords: Concept Drift; Adaptive Learning; Imbalanced data.

1 Introduction

In the real world data change over time. In many occasions it is assumed that the process generating an information stream behaves in a stationary fashion, but in most practical situations such an assumption does not hold. The data generation process at hand is often affected by unexpected, subtle changes – e.g. a measurement error of a temperature sensor – that may unchain a statistical decoupling

© Springer International Publishing AG 2017
C. Badica et al. (eds.), *Intelligent Distributed Computing X*,
Studies in Computational Intelligence 678, DOI 10.1007/978-3-319-48829-5_23

between the recorded data and the monitored parameter, which is commonly referred to as concept drift. From a more general perspective the aforementioned non-stationary behavior and the subsequent concept drift can be also unchained by seasonality or periodicity effects. Indeed many processes belonging to different application domains are driven by non-stationary data streams often subject to concept drifts, such as user modeling for personal assistance (e.g. spam filtering, personalized web and media search) when changing user interests, customer segmentation based on their preferences and needs, bankruptcy predictions and antibiotic resistance, among many others. As a result of concept drift predictive knowledge trained in the past becomes less accurate, thus recent developments in concept drift have heightened the need for algorithms and mechanisms for continuous adaptation in evolving environments.

Concept drift can be mathematically defined by denoting the Bayesian posterior probability a given data instance $\mathbf{x}$ belongs to class c as

$$P_{C|\mathbf{X}}(c|\mathbf{x}) \doteq P_{\mathbf{X}|C}(\mathbf{x}|c)P_C(c)/P_{\mathbf{X}}(\mathbf{x}), \tag{1}$$

where it is implicitly assumed that none of the above probability distributions changes over time. Concept drift arises when in any scenario the posterior probability $P_{C|\mathbf{X}}(c|\mathbf{x})$ does vary over time, i.e. $P_{C(t+1)|\mathbf{X}(t+1)}(c|\mathbf{x}) \neq P_{C(t)|\mathbf{X}(t)}(c|\mathbf{x})$. Considering this definition, the problem of learning in non-stationary environments can be studied from different perspectives, as surveyed in [1,2,3], such as *real* vs *virtual* drift, *sudden/abrupt* vs *incremental* drift, *gradual* vs *recurrent*, *active* vs *passive* approach, and *single* vs *ensemble* classifier.

Taken the above perspectives into consideration, it is important to point out the importance of the so-called stability-plasticity dilemma [4], which dictates the behavior and the results of a model prediction, and which refers to the ability of a model to retain existing knowledge or to learn new knowledge, but not being able to do both at the same time equally well. Most algorithms should take this dilemma as a reference when deciding the subset of instances (in case of windowing or reservoir sampling) or how many models (in case of ensembles) must be kept so as to predict the class for the next stream data. On this issue arises the necessity of considering how much storage and memory to allocate when facing a concept drift problem.

While the concept drift is itself challenging in non-stationary environments, a large number of concept drifting data sources also suffer from class imbalance, which refers to the case when the number of instances from different classes disproportionately differ from each other. Class imbalance is a well-known problem in the broad field of machine learning, but despite its widespread occurrence it has received much less attention in non-stationary data environments. It is often the case when these two challenges must be tackled at the same time, which implies the development of more sophisticated techniques blending together elements from both non-stationary concept learning and class imbalance. To this end, recent contributions gravitating on concept drift with imbalanced data aim at 1) detecting the drift as soon as possible; 2) adapting to the change quickly; and 3) recovering the accuracy level attained prior to the concept change.

This manuscript joins the relative scarcity of contributions dealing with the above twofold casuistry by presenting a radically new passive approach to handle concept drift, which we will henceforth label as PROSAMA.ID (PRObabilistic SAmple MAtchmaking for Imbalanced Data). The proposed scheme creates a set of new balanced population sets based on the most representative past instances, the predictive *essence* historically extracted from prior data. These essential past instances are combined together with data from the current batch and fed to an ensemble of models, whose training and subsequent voting permit adapting to sudden and recurrent changes quickly and renders a fast recovery rate of the prediction score right after the drift. When applied over a standard data set widely utilized for concept drift testing (namely, the SEA data set [5]), our proposed strategy is shown to score good average performance figures in these two algorithmic challenges, comparatively in the order of recent techniques from the state of the art (Learn++.CDS and Learn++.NSE [1]), but at a significantly reduced computational complexity and storage requirements.

The rest of the paper is organized as follows: Section 2 introduces the existing work in the field of imbalanced concept drift. Section 3 delves into the technical details of the proposed PROSAMA.ID approach, whereas Section 4 presents and discusses the simulation results obtained over the SEA data set. Finally, Section 5 draws some conclusions and outlines future research work.

2 Related Work

The change in data distribution for non stationary environments was first introduced by [6]. Years later, the problem of concept drift started to be increasingly popular when [7] presented his survey on concept drift, taking a special relevance nowadays as more and more information is organized in data streams rather than in static repositories; the fact that data distributions stay stable over a long period of time is far from realistic. Thereafter several recent reviews related to drift-aware learning were made available to the community in [1,2,8,9]. For the sake of conciseness we will just comment on recently reported techniques for adaptive learning in concept drift that have been proposed to deal with imbalanced data sets.

In what refers to recent passive ensemble-based approaches there are several batch-based learning algorithms that are gaining popularity in the last couple of years. As to mention, the Learn++ family is composed by Learn++.NSE [10], which is an algorithm for incremental learning in non-stationary environments; Learn++.CDS, which we will comment on later in this section; and Learn++.NIE [11], which is the naïve Learn++ scheme adapted for tackling non-stationary, imbalanced environments. Both Learn++ methods are based on ensemble learning, by which a new classifier is trained with each incoming chunk of data and added to the ensemble. It should be mentioned here the comparison between methods belonging to the Learn++ family that was posed in [12], where the authors provide a set of guidelines when deciding which is the best method: —Learn++.NSE— when the classes are relatively balanced, —Learn++.NIE—

when a strong balanced performance is needed on both minority and majority classes, and —Learn++.CDS— when classes contain imbalance and memory and computational complexity considerations are critical. Another remarkable ensemble technique, called Diversity for Dealing with Drifts (DDD), is the technical basis of the work presented in [13]. It consists of two ensembles: a low-diversity ensemble and a high-diversity ensemble. Both of them are trained with incoming examples, but only the low diversity ensemble is used for predicting. DDD always performed at least as well as other drift handling approaches. In [14] the authors present a framework called Ensemble of Subset Online Sequential Extreme Learning Machines (ESOS-ELM), which comprises 1) a main ensemble representing short-term memory; 2) an information storage module standing for long-term memory; and 3) a change detection mechanism to promptly detect concept drifts. Skew-Sensitive Boolean combination (SSBC) is an active approach presented in [15] that estimates target versus non-target proportions periodically during operations using the Hellinger distance, and adapts its ensemble fusion function to operational class imbalance. The basic idea of [16] is that when no concept-drift is present, their method (Selectively Re-train Approach Based on Clustering, SRABC) uses the most up-to-date chunk to train a base classifier and exploits it to update the ensemble classifier. This strategy improves the accuracy of the ensemble classifier because the most up-to-date chunk contains information about the current target concepts. When concept drifting occurs they use these data points with the current target concepts to re-train those base classifiers in the previous ensemble classifier. In this manner the algorithm efficiently converges to target concepts with high accuracy. The Drift Detection Method for Online Class Imbalance problems (DDM-OCI) is the method presented in [17], which attempts at reducing the minority class recall to effectively sense and capture the drift, responding to the new concept faster than their DDM-based counterparts.

While learning from a non-stationary data stream is by itself challenging, additional constraints such as class imbalance can make the problem even more difficult. Lately this constraint has received more attention with Learn++.CDS (Concept Drift with SMOTE), which is a batch-based algorithm that does not require any access to any historical data, but do require the pool of already trained models. It complements Learn++.NSE with SMOTE (Synthetic Minority Over-sampling TEchnique) [18], a combination of minority class over-sampling and majority class under-sampling that achieves better classifier performance than the mere under-sampling of the majority class. By virtue of this added functionality, Learn++.CDS has rendered good performance levels in synthetic and real data sets [12].

To end with we underline the importance of deciding which metrics are more suitable for the evaluation of an algorithm in an imbalance domain. In this regard it has been widely acknowledged in [2,9,16,19,20] that the Area under the ROC Curve (AUC) is the most strongly recommended score due to the fact that the naïve accuracy metric is not a reliable indicator in severely imbalanced data sets. This work will adopt this recommendation in what follows and specially in Section 4 for comparing results among different schemes.

3 Proposed Approach

Following the general schematic diagram depicted in Figure 1, we define as $\mathbf{X}_{tr}(t)$ and $\mathbf{C}_{tr}(t)$ the training batch at time t corresponding to features and the class to which each of the samples in the batch belongs, respectively. Likewise, $\mathbf{X}_{tst}(t)$ and $\mathbf{C}_{tst}(t)$ will stand for the test batch, whose it should be made clear that at time t only $\mathbf{X}_{tr}(t)$, $\mathbf{C}_{tr}(t)$ and $\mathbf{X}_{tst}(t)$ are available, which should be exploited jointly with the knowledge of past time instants $\{1, \ldots, t-1\}$ so as to predict $\mathbf{C}_{tst}(t)$ as accurately as possible. As argued above the prediction accuracy at each batch will be measured in terms of its AUC score, which will be denoted as $AUC(t)$. The novel approach proposed in this manuscript essentially creates new balanced populations for training a Z-sized ensemble of weak learners $\{M_t^z\}_{z=1}^Z$ based on the most representative past data instances and the training data of the present batch, as shown in Figure 2.

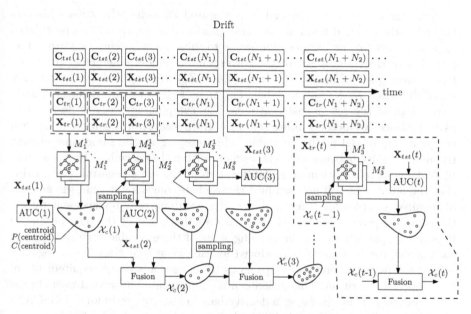

Fig. 1. Schematic diagram of the proposed PROSAMA.ID scheme.

The past predictive information utilized for building these training populations is stored in the form of *centroids*, i.e. a set of artificial data samples that represent a number of past training samples that are collectively predicted as belonging to the same class. The specific method to extract these centroids from the models trained in the past may vary depending on the particular predictive technique used for their implementation. For instance, in experiments subsequently discussed in Section 4 weak learners are instantiated as naive CART (Classification and Regression Trees), from which centroids can be inferred by traversing through the branches of the trained tree (3). The probability or relative weight

of each centroid (P(centroid) as shown in Figure 3) can be approximated by
the fraction of training samples that fall along the branch corresponding to the
centroid. Finally, the class of each centroid is the one to which the majority
of its represented samples belong. Other alternative schemes for producing this
centroid-based past predictive information may hold such as e.g. support vectors
and distance-based criteria in Support Vector Classifiers.

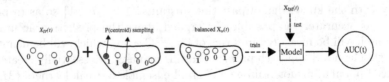

Fig. 2. Balanced training of the proposed PROSAMA.ID scheme.

The centroid extraction procedure is repeated for each of the Z weak learners
trained in the batch at hand, whose set of produced centroids $\mathcal{X}_c(t)$ is postulated
to provide valuable predictive information to subsequent batches by virtue of its
inclusion in the training set of the corresponding model ensembles. For batch
t the training set of model M_t^z ($z \in \{1, \ldots, Z\}$) is composed by the training
set of the batch $\mathbf{X}_{tr}(t)$ and its labels $\mathbf{C}_{tr}(t)$, as well as by an intelligently sam-
pled subset of the set of past centroids. The sampling strategy is designed so
as to produce balanced training sets of a given size S_{tr} (in samples) for each
compounding model of the ensemble, while keeping at the same time enough
training diversity between different learners within the ensemble. This is accom-
plished by sampling without replacement from the set of centroids. Provided
that the number of still non-sampled centroids becomes lower than the amount
of required samples for constructing a given training set, replacement during
the sampling process is allowed. This ensures that most of the past predictive
information participates in the training stage of the ensemble corresponding to
the current batch. Centroids produced at each time step t are fused together
with those from the previous steps so as to lessen the storage requirements of
the proposed algorithm. This fusion can be implemented in several ways; in the
experimental part of this paper a density-based clustering technique (DBSCAN)
[21] will be utilized, with probabilities being fused in an additive fashion.

The sampling procedure is done in a probabilistic fashion taking into account
three different factors: 1) the relative weight of the centroid with respect to the
model from which it was extracted; 2) a measure of *predictive similarity* of the
current batch to the one to which the model of the centroid belongs. This measure
can be generally formulated as a mapping function $\lambda_t : \{1, \ldots, t-1\} \mapsto \mathbb{R}[0,1]$
such that $\lambda_t(t') = 1$ if batches t and t' are *predictively similar* to each other
and $\lambda_t(t') = 0$ if they differ absolutely. The main rationale of this mapping is
twofold: on the one hand, the algorithm should select those past batches whose
training data resemble the current batch. On the other hand, similarity should
be valued jointly with the AUC performance of the algorithm at each batch.
Accordingly, $\lambda_t(t')$ is proposed to stem from the product between the similarity

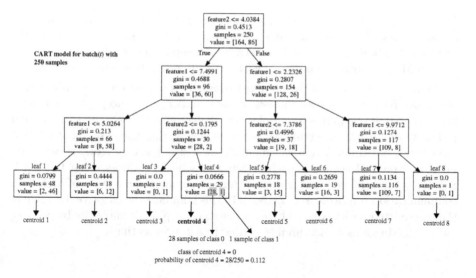

Fig. 3. Centroids extraction process for each batch.

of the imbalance ratios of batches t and t' and the AUC scores attained over the test set at batch t'. The computation of this measure over $t' \in \{1, \ldots, t-1\}$ and its sorting gives rise to an ordered list of predictively good past batches for the current instant t, which is used to drive probabilistically its centroid sampling.

4 Experimental Study

The performance of the proposed PROSAMA.ID approach has been evaluated when applied over one of the most widely used synthetic data sets for assessing new concept drift developments: the SEA data set [5]. Following the original data set generation procedure posed in this work, a total of 10000 3-dimensional samples have been generated at random within the range $\mathbb{R}[0, 10)$. Only the first two dimensions (features) are set informative for the class to be predicted, whereas the remaining dimension is irrelevant and acts as a noisy component for the target label. Points have been split in 100 batches of length 100 samples, which have been further divided into 4 main groups or blocks characterized by different concepts: a data sample belongs to class 1 if $x_1 + x_2 \leq \Theta$, where x_1 and x_2 represent the first two features of the sample and Θ is a threshold value that sets the frontier between the two classes. A recurrent series of values (i.e. $\Theta = \{4, 7, 4, 7\}$) has been used to generate the four concept blocks. An additional class noise has been also inserted within each block by randomly changing the class of 5% of the total instances. Based on these parameters PROSAMA.ID is applied over a total of 100 data batches, with $\mathbf{X}_{tr}(t)$ and $\mathbf{X}_{tst}(t)$ being composed by 100 samples. Every 25 batches a concept drift occurs, with 4 drift events in total. All weak learners are CART with maximum tree depth equal to 3, with ensembles of size $Z = 20$. As has been reviewed in Section 2, the family of

Learn++ algorithms have attained a remarkable performance over non-stationary environments. Therefore the goal of our experiments will be to compare the results of PROSAMA.ID to those by Learn++.NSE and Learn++.CDS when averaged over 10 Monte Carlo executions of each algorithm over the data set at hand.

Before commenting on the obtained results it is important to point out that the above parameter set establishes a quite challenging concept drift scenario: the small batch size makes it complex to balance properly between stability and plasticity due to the inherent risk of the models within the ensemble to overfit the current batch data. However a small batch size is a very realistic condition in many practical scenarios such as wireless sensor networks using the IEEE 802.15.4 physical layer specification, which is common for a suite of high-level communication protocols (e.g. Zigbee). In this standard the effective data size of the transmitted package is 133 bytes, whereas concept drift may arise by possible faults affecting sensor measurements (due to e.g. overheating).

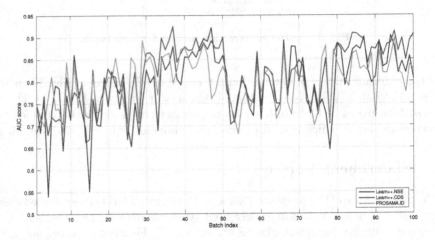

Fig. 4. AUC of PROSAMA.ID, Learn++.NSE and Learn++.CDS over the SEA data set.

As shown in Figure 4, the AUC performance of Learn++.NSE, Learn++.CDS and PROSAMA.ID when averaged first over all batches for a given Monte Carlo experiment, and second over all Monte Carlo experiments, is respectively 0.737, 0.746 and 0.764 for the batches belonging to the first block ($\Theta = 4$). Due to the recurrent concept drift, in the third block all algorithms should learn from the first block. So do the algorithms in the benchmark, as the impact of the concept drift on their AUC score results to be minimal. When averaged over the entire data set and all the Monte Carlo executions, the average AUC scores are 0.782 (Learn++.NSE), 0.815 (Learn++.CDS) and 0.798 (PROSAMA.ID). Although the proposed approach does not render the best average results within the benchmark, it is fair to remark that it overrides the need for storing past models and evaluating their performance when predicting the current batch. In fact it can be shown that PROSAMA.ID requires, for the parameter configuration tested in this paper, less than 50% of the total model evaluations required by its `Learn++` counterparts.

5 Conclusions and Future Work

Learning in non-stationary environments represents a challenging and promising area of research in machine learning; the increasing relevance of concept drift in real-world applications has increased with the proliferation of streaming and Big Data applications. In those scenarios traditional data mining approaches that ignore the underlying drift are inevitably bound to fail, hence demanding effective solutions that can adapt to concept changes. This research work has elaborated on a new approach, coined as `PROSAMA.ID`, which has been shown to be innovative with respect to the state of the art when using partial old data (centroids) to generate new training data, and to yield a promising behavior when adapting to abrupt and recurrent concept drifts. When dealing with data streams in non-stationary environments one must have in mind that such an scenario imposes additional constraints in terms of processing time, memory usage or recovery rate. The proposed approach is an attempt at reducing the amount of past information required to build the models, and proposing just the storage of representative past instances (centroids). Experimental results have suggested that `PROSAMA.ID` could be competitive with `Learn++` algorithms, but without the need for storing all the past trained models.

Future efforts will be devoted to the validation of the innovative technical approach featured by `PROSAMA.ID` when applied over other synthetic and real data sets from the literature, as well as on real scenarios characterized by gradual and/or incremental drifts. Finally, aspects such as limited time or computational cost will be considered, possibly by adopting strategies for detecting data obsolescence.

6 Acknowledgements

This work was supported by the Basque Government under its ELKARTEK research program (ref: KK-2015/0000080, BID3A project).

References

1. Ditzler, G., Roveri, M., Alippi, C., Polikar, R.: Learning in Nonstationary Environments: A Survey. IEEE Comp. Int. Magazine, 10(4), 12–25 (2015)
2. Žliobaitė, I., Pechenizkiy, M., Gama, J.: An Overview of Concept Drift Applications. Big Data Analysis: New Algorithms for a New Society, 91–114 (2016)
3. Hoens, T. R., Polikar, R., Chawla, N. V.: Learning from Streaming Data with Concept Drift and Imbalance: an Overview. Progress in Artificial Intelligence, 1(1), 89–101 (2012)
4. Grossberg, S.: Nonlinear Neural Networks: Principles, Mechanisms, and Architectures. Neural Networks, 1(1), 17–61 (1988)

5. Nick Street, W., Kim, Y.: A Streaming Ensemble Algorithm (SEA) for Large-Scale Classification. ACM SIGKDD International Conference on Knowledge Discovery and Data Mining, 377–382 (2001)

6. Schlimmer, J. C., Granger, R. H.: Incremental Learning from Noisy Data. Machine Learning, 1(3), 317–354 (1986)

7. Tsymbal, A.: The Problem of Concept Drift: Definitions and Related Work. Computer Science Department, Trinity College Dublin, 106:2 (2004)

8. Heywood, M. I.: Evolutionary Model Building under Streaming Data for Classification Tasks: Opportunities and Challenges. Genetic Programming and Evolvable Machines, 16(3), 283–326 (2015)

9. Gama, J., Žliobaitė, I., Bifet, A., Pechenizkiy, M., Bouchachia, A.: A Survey on Concept Drift Adaptation. ACM Computing Surveys, 46(4), 44 (2014)

10. Elwell, R., Polikar, R.: Incremental Learning of Concept Drift in Nonstationary Environments. IEEE Transactions on Neural Networks, 22(10), 1517–1531 (2011)

11. Ditzler, G., Polikar, R.: An Ensemble based Incremental Learning Framework for Concept Drift and Class Imbalance. International Joint Conference on Neural Networks, 1–8 (2010)

12. Ditzler, G., Polikar, R.: Incremental Learning of Concept Drift from Streaming Imbalanced Data. IEEE Transactions on Knowledge and Data Engineering, 25(10), 2283–2301 (2013)

13. Minku, L. L., Yao, X.: DDD: A New Ensemble Approach for Dealing with Concept Drift. IEEE Transactions on Knowledge and Data Engineering, 24(4), 619–633 (2012)

14. Mirza, B., Lin, Z., Liu, N.: Ensemble of Subset Online Sequential Extreme Learning Machine for Class Imbalance and Concept Drift. Neurocomputing, 149(Part A), 316–329 (2015)

15. De La Torre, M., Granger, E., Sabourin, R., Gorodnichy, D. O.: Adaptive Skew-sensitive Ensembles for Face Recognition in Video Surveillance. Pattern Recognition, 48(11), 3385–3406 (2015)

16. Zhang, D., Shen, H., Hui, T., Li, Y., Wu, J., Sang, Y.: A Selectively Re-train Approach based on Clustering to Classify Concept-Drifting Data Streams with Skewed Distribution. Advances in Knowledge Discovery and Data Mining, 413–424 (2014)

17. Wang, S., Minku, L. L., Ghezzi, D., Caltabiano, D., Tino, P., Yao, X.: Concept Drift Detection for Online Class Imbalance Learning. International Joint Conference on Neural Networks, 1–10 (2013)

18. Chawla, N. V., Bowyer, K. W., Hall, L. O., Kegelmeyer, W. P.: SMOTE: Synthetic Minority Over-sampling Technique. Journal of Artificial Intelligence Research, 16, 321–357 (2002)

19. He, H., Chen, S.: Towards Incremental Learning of Nonstationary Imbalanced Data Stream: A Multiple Selectively Recursive Approach. Evolving Systems, 2(1), 35–50 (2011)

20. Ditzler, G., Polikar, R., Chawla, N. V.: An Incremental Learning Algorithm for Non-stationary Environments and Class Imbalance. International Conference on Pattern Recognition, 2997–3000 (2010)

21. Ester, M., Kriegel, H. P., Sander, J., Xu, X.: A Density-based Algorithm for Discovering Clusters in Large Spatial Databases with Noise. KDD, 96(34), 226–231 (1996)

Author Index

© Springer International Publishing AG 2017
C. Badica et al. (eds.), *Intelligent Distributed Computing X*,
Studies in Computational Intelligence 678, DOI 10.1007/978-3-319-48829-5

Printed in the United States
By Bookmasters